Thomas Bryant

The Surgical Diseases of Children

Thomas Bryant

The Surgical Diseases of Children

ISBN/EAN: 9783337035075

Printed in Europe, USA, Canada, Australia, Japan

Cover: Foto ©berggeist007 / pixelio.de

More available books at **www.hansebooks.com**

THE
SURGICAL DISEASES

OF

CHILDREN.

BEING
THE LETTSOMIAN LECTURES DELIVERED BEFORE
THE MEDICAL SOCIETY OF LONDON,
MARCH, 1863.

By THOMAS BRYANT, F.R.C.S.,

ASSISTANT-SURGEON TO GUY'S HOSPITAL.

LONDON:
CHURCHILL & SONS, NEW BURLINGTON STREET.
MDCCCLXIII.

LEWES:
GEO. P. BACON, PRINTER.

TO THE

PRESIDENT, VICE-PRESIDENTS,

AND FELLOWS

OF

THE MEDICAL SOCIETY OF LONDON,

THE AUTHOR

RESPECTFULLY DEDICATES

THE

FOLLOWING PAGES.

TABLE OF CASES REFERRED TO IN THESE LECTURES.

	PAGE
I.—Imperforate anus	34
II.— ,, rectum	40
III.— ,, anus, with recto-vaginal fistula	48
IV.—Complete absence of perinœum	49
V.—Absence of fibula, os calcis, cuboid, and three metatarsal bones, as well as the toes	53
VI.—Concussion of the brain	59
VII.—Tetanus (traumatic)	63
Ditto ditto	,,
VIII.—Necrosis of frontal bone	64
IX.—Laceration of lung tissue without fractured ribs	74
X.—Salivary Fistula	81
XI.—Ranula	82
XII.—Warty growths upon the tongue	83
XIII.—Wound of tongue	85
XIV.—Enormously enlarged tonsils	86
XV.—Polypus recti (three cases)	89
XVI.—Irritable bladder (four cases) due to state of prepuce and retained secretion	93
XVII.—Cases of inflammation of the pulpy layer between the shafts of long bones and their epiphyses	114
XVIII.—Intra-uterine fracture	117
XIX.—Green-stick fracture	120
XX.—Necrosis in shoulder-joint, and recovery with anchylosis	130
XXI.—Necrosis in hip-joint—operation and recovery	131
XXII.—Compound dislocation of ankle-joint—perfect recovery	132
XXIII.—Extensive disease of ankle-joint—thorough recovery by rest	133
XXIV.—Congenital fatty tumour	137
XXV.—Pedunculated fibro-cellular tumour	138
XXVI.—Simple cystic tumours	139
XXVII.—Keloid	140
XXVIII.—Tumours of testis, simple and malignant	141

PREFACE.

THE following Lectures have already appeared in the pages of one of the weekly medical journals; and it is owing to the kind reception with which they have been favoured, and the solicitations of friends, that the author has been induced to re-print them in the present form. They are published simply as contributions to the subjects upon which they treat, with the hope that such facts and opinions as they contain may not only be read with interest, but may at the same time tend to advance the sum of our professional knowledge.

2, *Finsbury Square,*
September, 1863.

THE

SURGICAL DISEASES OF CHILDREN.

LECTURE I.

Mr. President and Gentlemen,—The diseases of children have, from an early period of the practice of the healing art, monopolised a considerable portion of the attention of the surgeon; although it has been reserved for men of the present century to enter more scientifically into the subject, to search out the peculiarities of the child's diseases, and to elucidate the differences which are to be observed between the affections of early and adult life. Hence it is only within the last few years that the diseases of children have been shaped into a speciality, and have exclusively occupied the attention of some of the leading members of our profession.

It can be scarcely doubted that we are indebted for this sudden increase in our knowledge of such an important branch of medical and surgical study, to the rapid advance of physiology as a science, for how could men comprehend the differences between the diseases of childhood and of adult age, when they were comparatively ignorant of the processes by which health was maintained, and more particularly of the means by which the development and growth of the infant structure passed on to the perfection of mature life?

To this advance in physiological study may be fairly attributed our more correct knowledge of pathology: and it is upon this solid and sound basis that our present improved acquaintance with the diseases of early life has been, unquestionably, reared.

The Council of this old and honoured Society, recognising the importance of this subject, have been led to believe that the interest of its members might be promoted and the profession benefited by having their attention drawn from the broad field of general medicine and surgery to the comparatively small one occupied by that of the diseases of children: and, on the strength of this belief, have, abandoning the custom by which they have been hitherto bound, defined the subjects for the Lettsomian Lectures, before appointing the lecturers selected for their delivery. In carrying out these intentions, the Council have nominated me to deliver their surgical course; and they will not, therefore, be surprised that I should confine my attention exclusively to the surgical aspect of the subject; they well knew, also, that my attention had not been more particularly devoted to the diseases of early life than to those of the adult; and I therefore take it that they believed the subject would be most advantageously illustrated by one whose experience, embracing the broad field of surgery, had not been narrowed by the contracted view too often engendered by special practice in any one class of medical or surgical diseases.

I propose, therefore, in carrying out these assumed intentions of your executive body, to occupy your attention during these three Lectures, by more particularly pointing out the differences which are to be found between the surgical affections of early and adult life. Having, in the first place, glanced at the differences in the physiology between the two classes, and shown how the pathological processes are modified by such conditions, I shall lastly endeavour to point out in what way this knowledge should influence our practice both in the treatment of disease and injury. I shall review the subject in this order, taking up the various systems in rotation, after having dwelt, as far as time will allow, upon some of the special surgical affections of early life.

THE DIFFERENCES BETWEEN THE PHYSIOLOGY OF THE ADULT AND OF THE CHILD.

I doubt whether I shall be deemed to be departing far from the truth when I assume it as a physiological fact, which all men readily admit: That in *adult life* the vital forces, started, supported, and maintained by the nervous, respiratory, circulatory, and glandular systems, are mainly employed for the simple preservation of man's physical structure: or, in other words, for the maintenance of what has already attained its perfect growth and complete development. I do not deny that these forces may exist for other subsidiary, or perhaps higher purposes; but it is for the maintenance of the body that they are doubtless principally concerned; and for the present, this fact alone is one which I would have you bear continually in mind during the consideration of the subject upon which we are now engaged.

The second point, which I would also wish to be impressed with equal force upon your minds, and which I believe will be accepted as readily as the former, may be thus aphoristically expressed: That in *infant and child life*, the vital forces, originated, supported, and maintained by the nervous, respiratory, circulatory, and glandular systems, are mainly directed for the purposes of growth and development of the being's physical structure; a portion only of their power being employed for its preservation and maintenance.

The grand distinction between the two classes rests upon the fact that in the child the vital forces are directed for its *growth and development*, whilst in the adult they are mainly employed for the *maintenance* of man's physical structure.

Under these apparently diverse physiological conditions, it can excite no surprise that there should exist a material difference between the diseases of early and of adult life, and that the pathological processes should be somewhat modified by the physiological principles. For in man's high nature, in which the physical forces are so wonderfully correlated, the influence of growth and development upon *diseased* processes is too great to be passed over without attention; for it is to the abnormal direction given to these *natural* processes that most of the special affections of infancy are to be attributed, and in no single morbid action in young life can their influence even for one moment be disregarded. Again, the forces which are employed in the building up of the body, and in maintaining it when completed, may be merely varying expressions of the same power applied in different directions; yet, growth, development, and maintenance are so distinct in their several natures, and so uniform in their ends, that it might be excusable if, for a moment only, we were to regard them as originating from independent sources, and as expressions of distinct powers; for daily practice proves to us that the growth of a part may go on in all its completeness, whilst its development may be arrested, or progress in an abnormal direction; or we may witness the development progressing in its normal channel, with growth showing itself either in excess or in diminished force; and yet during this time the maintenance of the whole body may be preserved. All malformations are, therefore, but results of some defective power, either in the process of growth or of development, or of both; and they are always to be explained by the preponderance, diminution, or abnormal manifestation of one or of the other. Harelip, fissured palate, monsters, cerebral tumours, and spina bifida, etc., are to be explained by some defect in the natural development of the child; its growth and maintenance being, in the majority of instances, perfect of its kind; whilst instances may presently have to be recorded, in which growth alone has proved defective, development being perfect.

THE GREATER ACTIVITY OF INFANTILE DISEASES.

We will now pass on to the consideration of another distinctive feature of infantile diseases, which characterises one and all; and that is, their greater activity when compared with those of adult life.

This pathological fact is doubtless to be explained by another physiological truth, which is applicable to adult as well as to child life, although it may be more particularly exemplified in the last, viz.:

That the activity of disease appears to be in direct proportion to the activity of the vital processes going on in a part; that in highly vascular organs morbid processes progress more rapidly than they do in others less favoured; and that in proportion to the extent of action going on in a part will the development of morbid phenomena be manifested when once originated.

In early life, therefore, the well known activity of the vital processes in all the organs renders them particularly prone to morbid phenomena; and those parts in which development is progressing the most actively are the most liable to become the seat of morbid action.

Again: During development and growth, cell-life is necessarily active; for it is needless to dwell upon the fact that it is through such cell-growth that all development and increase must be carried out. In morbid actions, therefore, when once originated, a like cell-growth and multiplication is to be observed; and this point is well exemplified in the case of tumours. For such tumours, when once developed in children, are generally of rapid growth; and, whether simple or malignant, their peculiar characteristic is cell-structure; the simple tumours being formed by a rapid repetition of their original elements; and the malignant consisting almost entirely of a cell-growth invariably of the so-called medullary character; the vital processes manifesting their power by the rapid growth and repetition of the original cell, which had gone astray and become a prodigal, in lieu of developing into a structure of a more perfectly organized nature. Their growth, also, is generally rapid, being in accordance with the activity of the other functions; healthy and diseased processes having, in this respect, a strong analogy.

As I proceed with my subject, I shall have the pleasure of illustrating more fully the truth of these remarks; and I propose, therefore, to pass on to discuss more particularly the specialities of the surgical diseases of children, commencing with such as are to be explained by some defective power either in the growth or development of the infant structure,—the defects which are to be found in the digestive system having our primary attention.

HARE-LIP.

Commencing at the superior orifice of the mucous canal which forms part of the digestive system, the deformity denominated hare-lip is the most important surgical affection that demands our notice; and it is one which, without doubt, is to be explained by a distinct want, or arrest of development of the fœtal structure.

Anatomists are now generally satisfied that the upper jaw is developed from several centres; and that its incisive portion is the analogue of the permanent intermaxillary bone of the lower animals. The development of the integument forming the lips is also modified, or is rather determined, by the development of its bony attachment; and it is, therefore, by such an arrest of this process that a hare-lip, associated or not with a fissured alveolar process or a divided hard or soft palate, is to be explained.

A hare-lip is, therefore, a simple want of union between the tissues which are developed from different centres; and a fissured palate, partial or complete, is but a want of union between the osseous centre—the analogue of the intermaxillary bone of the lower animals—and the body of the superior maxillary bone.

The fissure in the soft parts or in the bone may, consequently, be either single or double, partial or complete; the extent of the deformity being determined by the period at which the arrest of the development of the parts was primarily fixed, or by the mal-development being confined to the junction of one or of both intermaxillary bones.

I might add, in passing, that if some anatomists are doubtful as to the existence of this intermaxillary bone in the human species, on account of the difficulty, not to say impossibility, of separating it from the other osseous centres in a healthy or adult bone, pathological investigations may be advanced as powerful means of proving the truth of the theory; and the following case, quoted from my own experience, is one in point.

A child, aged 3 years, was brought to me at Guy's Hospital in the year 1858, with necrosis of some portion of the upper jaw, consequent on a severe attack of measles. With a pair of dressing forceps the dead bone was readily removed, and thorough convalescence followed. The bones which were removed turned out to be perfect specimens of the incisive centre of bone, or the intermaxillary; and they came away entire, fairly proving that they had been originally developed as independent centres of ossification. The bones were exhibited before the Pathological Society, in the year 1858.*

We will now return to the more practical points connected with this subject of hare-lip; and I believe that I shall be able to bring before your notice some facts which cannot but be regarded as of interest, adduced from the material which my opportunities at Guy's Hospital have enabled me to collect. I possess the records of sixty-four cases of this deformity; this number including all the examples which have been admitted into Guy's Hospital for about eight years, with other cases which have fallen under my own care amongst the out-patients, these forming one-third of the whole number. A careful analysis of this material has yielded the following facts.

Nature and Extent of the Deformity. The arrest of development causing the deformity known as hare-lip may be of different degrees of severity. It may be confined to the lip itself, or extend to the hard and soft palate, causing a mere fissure in the gum, or a complete separation of the hard and soft tissues. It may be of a single or a double nature;

---

* I have since removed from a man another specimen equally perfect.

the double fissure, however, never extends beyond the lip and the intermaxillary bone, the cause of the deformity forming a sufficient explanation of this fact.

I have seen, in the lip of an adult, a congenital cicatrix on the left side of the upper lip, with a slight elevation of its mucous margin, evidently due to a condition which may be said to be the very earliest indication of a hare-lip. The child of this parent had a more complete fissure, and was operated upon by me with complete success.

The different varieties and complications, however, of hare-lip will be best shown by an analysis of the collected cases; and the following facts may be relied on.

In 32 of the 64 cases, the deformity was of a simple character, and affected the lips alone to a greater or less extent; 3 cases were complicated with a fissure of the gum, corresponding to the labial deformity; 2 had a fissured gum and hard palate; 21 a fissured hard and soft palate; and 6 cases were of a double nature.

It is thus evident that half the cases of hare-lip are of a simple and uncomplicated character, and that in the remaining moiety some deformity of the mouth and palate coexists; a fissure of both the hard and soft palate being present in two-thirds of the remainder, or in one-third of the whole number of cases.

Seat of the Deformity. All the cases which have passed under my observation have been in the upper lip; and, although authors refer with some confidence to the fact that the lower lip may be similarly affected, it is difficult to believe in the truth of such an opinion, as the very pathology, or rather the explanation of the nature of the deformity, forbids the idea of the possibility of such an occurrence holding any position in our minds; and, if the explanation which has been just given be a true one, the opinion that it is ever found in the lower lip must be looked upon as of doubtful accuracy.

The upper lip, therefore, is its natural seat; and the next point for illustration relates to the side on which it is usually observed. That it is never found in the median line, is a fact which the experience of all surgeons will at once verify; and the results of experience upon this point only bear out the truth of the explanation of its cause. That it is generally found on the left side of the body, the analysis of my own cases positively proves, for 63 per cent. of the cases were on the left side, and 36 only on the right. A satisfactory explanation of this fact has never yet been given, although I believe it to be a generally received truth that deformities, as a rule are more common on the left than on the right side of the body, and the above figures tend in a great measure to prove the accuracy of the remark.

Sex. The influence of sex is the next point which presses upon our attention; and, if the facts which the analysis of my materials yields, were not supported by the generally received impression which most surgeons entertain, they might be considered to be somewhat strange; for out of the 64 cases of hare-lip, 44 occurred in boys, and but 20 in girls, or in the proportion of 70 per cent. of the former, and 30 per cent. of the latter, this deformity occurring 40 per cent. more frequently in males than in the female sex.

If we analyze these cases again, however, an unexpected result makes its appearance; for in the simple and uncomplicated examples of hare-lip, the difference between the two sexes is not so great; as of the 30 cases, 18 or three-fifths were in males, and 12 or two-fifths in females. But in the more complicated examples—that is, in those in which the hard and soft palate were more or less involved—the greater frequency of its occurrence in the male sex becomes most apparent. Of the cases in which both hard and soft palate were completely fissured, 17 cases, or 80 per cent., were in boys, and but 4 cases, or 20 per cent., in girls, the male element predominating to an extreme degree, while in the worst cases of all—that is, of double hare-lip—all the six examples were observed in male children.

These facts are, to say the least, extremely curious; for it can hardly, I think, be looked upon as an accident that, in the cases which have passed under my own observation, the male element should have predominated so largely; and I am disposed to look upon my own experience as an expression of what would be that of others, if it were given to us. By way of summary, the following conclusions may be drawn.

Conclusions. 1. Hare-lip is always found in the upper lip, and is most frequent on the left side of the body; 63 per cent. of the cases taking place on the left side, and 36 per cent. on the right.

2. It is more common in the male than in the female sex, in the proportion of 70 per cent. of the former to 30 per cent. of the latter.

3. In the simple uncomplicated cases, the proportion is less marked; but as the cases become more complicated, the greater frequency of its occurrence in the male sex becomes more apparent, the proportion between the sexes being 80 per cent. and 20 per cent. It would also appear to be a rare thing to find a double hare-lip in a female child.

Treatment of Hare-Lip. There neither is nor can be any difference of opinion between surgeons as to the expediency of affording surgical relief in the cases of deformity which we are now considering; although there is still great uncertainty and diversity of opinion between them as to the period at which it is most advisable to submit the child to an operation. Some surgeons assert that operative interference is most successfully carried out at the very earliest period of the child's existence, and that the earlier the operation is performed, the greater are the chances of complete success. On the other hand, surgeons of equal eminence maintain that the practice just laid down is fraught with danger, that the correct practice is to wait till the child's powers are fairly established, and that one or even two years should be allowed to pass before any attempt to repair the deficiency be made. Between these two extremes lie the opinions of many men; and it is, therefore, with a view of obtaining some definite data upon which a positive opinion can be based, that I will now lay before you the conclusion to which I have been led.

But, before doing so, allow me to allude, with all deference, to the uncertain grounds upon which surgeons have been in the habit of founding positive opinions, for it is too true that these opinions are, as a rule, based only upon impressions, although it may be that these impressions are the result of extensive and perhaps most conscientious experience. Still I feel that I may appeal with confidence to all who hear me, when I assert that there are few things more fallacious than the general impressions which experience affords, however great may have been the field from which they have been gleaned; and that true impressions can only be gained from positive and definite materials.

I am unable to direct you to any work or even journal in which definite data have been recorded, from which just conclusions can be drawn as to the period at which it is most advisable to operate in cases of hare-lip, unless I were to mention my own previous contributions to this subject, published in the third part of my *Clinical Surgery*.

It is, therefore, with some confidence that I am now able to add something to our stock of knowledge which may prove of use in guiding us to a correct decision. If I lay myself open to the charge of saying that surgeons have hitherto formed their opinions upon indistinct and indefinite data, I can only justify myself by pleading that the grounds upon which the opinions were based have not been published; and that, therefore, I may be forgiven for doubting their existence.

The facts which I now adduce may form but a nucleus round which others may hereafter collect; and the conclusions I now give will have to be confirmed or corrected by future observers.

Analysis of Cases. Four cases were operated on within the first two weeks; one on the fifth day, and two on the eleventh, one proving fatal.

Eight cases were between four and five weeks of age; in two of them, the parts subsequently gave way, although the cases in the end turned out well.

Ten cases were operated on between the sixth and eleventh weeks, with success.

Fifteen cases were treated between the third and sixth month, with only one failure.

Six cases were operated on between the sixth and twelfth month, with a good result; and

Fifteen were also successfully treated after the first year.

From these facts it will appear that, during the first six weeks of life, the operation for hare-lip is by no means so successful as to warrant its general performance, unless an absolute necessity compels. Out of the twelve cases, one died; and in two the wound subsequently sloughed, although in both a good recovery was finally secured. At a later period of life, a successful issue was recorded in all.

At what period of the infant's life, then, should the operation for hare-lip be generally performed? In order that a satisfactory answer should be given to this question, it is necessary to consider the purposes for which the operation may be required; and they may be divided into two classes.

Firstly, the operation may be called for to preserve life; the imperfection of the mouth forming an insuperable difficulty to the child's sucking.

Secondly, it may be called for simply as an operation of expediency, to improve appearances and remove defects.

When demanded for the satisfaction of the first purpose—namely, to preserve life—there is no question of period which the surgeon has to decide. The operation must be undertaken at all hazards; but the successful termination of the case will be uncertain. For this purpose I have performed the operation on the fifth day with complete success, and other surgeons have been equally successful even at a much earlier period; but, under these conditions, the necessity of the operation overcomes all other questions, and the surgeon has no option but to do his best.

In the second and more numerous class of cases, however, the time for operation rests entirely on the surgeon's will; and it only remains for us to decide, from the records of positive experience, the period of life at which the greatest success can be anticipated.

Judging from the materials which I have just laid before you, it would appear that the operation when performed at any period of life after the sixth week is likely to be followed by a successful result; I would give the preference, however, to about the third month of life, the vital powers of the child having by that time become fairly established, and well able to resist the tax upon their strength which is necessarily occasioned by any operation. At an earlier date—that is, during the first few weeks of life—the operation should be condemned; unless, as has been previously explained, the existence of the child appear imperilled.

The Operation. It would be hardly profitable to any of us if I were to take up your time by recapitulating all the varieties of operation which have been suggested by various surgeons for the relief of the deformity now under consideration; and I propose, therefore, to confine my remarks simply to what I have observed and practised, and to recommend what appears to be the best means out of the many which have received attention.

In simple cases, there are but two main objects which the surgeon has in view—to pare the edges of the fissure and to adapt them so as to render the deformity as slight as possible.

I would give the preference to the scalpel for the performance of the first stage ; for with such an instrument a cleaner section of the tissues is made than when scissors are employed, and this cleanness of the incision is a point of primary importance.

The form of section which is to be recommended is variously estimated by different surgeons. Some are content with a clean straight section of the lip's margin, being so satisfied with obtaining simple union of the separated parts, as not to care about making an attempt to supply the deficiency of tissue always so noticeable at the margin of the lips. Other surgeons recommend that the line of incision should be curved inwards, so that the lower margin of the lips, when brought into position, will be made to project downwards, and thus tend to correct the evil which has been just alluded to. But I am disposed to give the preference to the plan of operation which I believe was originally suggested by Malgaigne, as by it the labial notch is almost to a certainty done away with, and the deformity most completely remedied. It consists simply in paring the edges of the wound from above downwards, leaving the inverted flap adhering to its labial border. When this is done, and the upper edges of the divided lips have been brought together, the lower flaps may be connected by a fine suture; and, if proved to be too long, they may be curtailed, sufficient material being left to fill in the gap which is too often the result of the other forms of operation. In my hands, this plan of operation has proved eminently successful, and it only requires in its application a little nicety in the adjustment of the parts.

An important preliminary point, however, demands attention, as the success of the case rests materially upon its due performance; for, however well the marginal incisions may be executed, unless the lip has been rendered readily moveable upon its labial attachments, a failure in the operation must be expected. This freedom of movement is, however, readily secured by freely separating the mucous membrane of the lip from its osseous connexions. When this is done, the whole lip can be readily raised from its position, and all chances of tension are completely taken away.

The fear of removing too much of the lip's margin is one which is most certainly groundless, the usual fault of operating surgeons lying the other way; for the labial tissues are very extensible, and a free section of the parts is to be preferred.

The *second step* of the operation remains for us to describe; its object being to bring the edges of the wound together, and to keep them there. This end is most readily and efficiently carried out by means of the interrupted suture. Silk or wire may be selected according to the fancy of the operator, one exciting as much or as little irritation as the other. The interrupted suture is to be preferred to the uninterrupted, as it is more readily removed.

The sutures should be inserted from a quarter to half an inch from the wound's margin, and carried obliquely through the lip, to the line which is bounded by the mucous membrane, but not through it, and then introduced at a like spot on the opposite side, and firmly tied. A fine suture at the labial margin should also be applied, in order to maintain as accurately as possible the line of lip.

All bleeding is generally at once arrested when the wound's margins are brought into contact; but, if it be troublesome, one of the sutures may be made to perforate the bleeding vessel.

A little elastic collodion may be subsequently painted over the part, to prevent friction; but, as a rule, the practice of leaving the part open proves equally satisfactory. The employment of pins in simple cases may certainly be looked upon as unnecessary, as the simple means which have been described are amply sufficient.

In operating, therefore, upon uncomplicated cases of hare-lip, the following points appear the most important.

First, to separate the lip freely from its gummy attachments; *secondly*, to make a free section of its edges, according to the plan previously laid down;

and *thirdly*, to bring the edges accurately together by fine interrupted sutures, introduced at a distance from the wound's margin, and deeply placed. If these points are observed, and the child is neither too young nor too sickly, a successful termination to the case may with some confidence be predicted.

Complicated Cases of Hare-lip. The remarks which I have just concluded, concerning simple and uncomplicated hare-lip, are equally applicable to other cases of a more complicated nature; but there is a point of practice in *double* hare-lip which deserves a few moments' attention, as there is still in such cases a diversity of practice, which, I confess, appears to me somewhat singular. Some surgeons invariably operate only upon one side at a time, fancying that a greater success follows such a plan than when both sides are treated at one operation. From what has passed under my observation, I cannot see any reason why both sides should not be simultaneously treated; and I further believe that advantages are obtained by such a proceeding, which are not secured by the more timid and prolonged process. In the six examples of double hare-lip which I have tabulated, primary success followed in each instance; and in the cases which fell into my own charge, I had no reason to believe that the operation would have succeeded better if a different practice had been followed. In one case, of a boy aged nine weeks, I employed a pin in order that the centre bit should be well held down to the lateral portions; and good success followed the attempt. In the second case, of a boy aged one month, I operated after the plan which I have previously given, but I preserved only the lateral flaps of the outer portions, joining these in the centre beneath the central piece. Primary union followed the operation, and a recovery in which the deformity was but little marked.

In no instance have I witnessed any evils resulting from the practice which I have just laid down; and I am at a loss to understand the principle upon which surgeons still adhere to the older practice of carrying out a double and separate operation.

In certain complicated cases of hare-lip, the central incisive or intermaxillary portion of bone is found projecting to a detrimental extent; and it becomes a question with the surgeon whether he shall remove it by means of bone-forceps, or fracture it and press it backwards. In many instances, unless some such proceedings as these be carried out, the operation for hare-lip cannot be performed. I have seen both plans executed with good success in the practice of my colleagues, and in one instance have performed excision with success.

If the bone can be pressed backwards by a moderate amount of force, the practice appears to be preferable to the one in which the obstacle is only overcome by its removal; but if this plan should be impossible or inexpedient, there is no alternative but to excise the part. It is true that, by removing the central portion, the incisor teeth will be taken away; but it is better that such a practice should be carried out, than that the original deformity should be left, particularly when it is remembered that, with such a projecting bone, the incisor teeth would in all probability appear much out of place, even to the extent of piercing the lip—a contingency which I have in one case certainly observed.

In some instances, however, an exfoliation of the incisor tooth may take place before any operative interference has been performed; and this was illustrated in a case which fell under my care on Oct. 27th, 1862, in a male child aged one month, who had a hare-lip on the left side, with a fissured hard and soft palate, the incisive portion of the superior maxillary bone projecting far forwards. When the mother brought the child to me, the soft parts over the projecting bone were dry and apparently dead; and with the slightest touch the "scab" came off, with it also the crown of the incisor tooth. The surface readily healed, and the child is to return to me for operation at a future period.

There is one other point in the treatment of these cases which the surgeon should remember, viz., that if the first operation fail, and union by first intention cannot be procured, there is still a good hope of securing ultimate success by union from secondary adhesion. This end is to be obtained by the application of strapping, or, in preference, sutures, to the granulating wound, so that the edges may again be placed in apposition. In several instances which have been tabulated, this practice was followed, and a good result took place; and in two cases which came into my hands several years ago, equal success can be recorded. Both were infants, upon whom I had been led to operate at a very early period, two and three weeks only being their respective ages. Sloughing followed the original operation; and, when the parts had begun to granulate, sutures were re-applied, and a good recovery was eventually secured.

With these remarks I will now conclude the subject of hare-lip, having, I believe, touched upon the chief points of practical importance, and given facts upon which definite opinions may be based. The personal experience of any single surgeon in hare-lip is rarely very great, and it can only be on the collective experience of many that positive opinions can be formed. I advance my own facts, knowing them to be true; and entertain the hope that other surgeons may be induced to add their own stock of cases to these which I now contribute, and thus either confirm or correct the opinions which have been here put forward.

The subject of cleft palate is not one which need occupy our attention; for, although it is a congenital defect, it can be remedied only in advanced life, and has therefore no special claim for notice. Atresia Oris, or closure of the mouth, is also a congenital condition which may exist; but as I have had no personal experience upon the subject, it will receive only this passing notice.

MALFORMATIONS OF THE ANUS AND LOWER BOWEL.

Having thus far occupied your attention by dwelling for some length upon the congenital deficiencies which are to be observed at the upper orifice of the mucous or digestive canal, I propose to pass downwards to the other or terminal end, and discuss the subject of imperforate anus or occluded rectum; and to attempt, from general as well as from personal experience, to bring before you some practical points, which may be of service in the treatment of this interesting and difficult class of cases. These malformations of the rectum may be divided into three great sections.

1. The simplest form, which includes the cases in which the orifice of the anus is completely closed, the rectum being partially or wholly deficient.

2. The more complicated class of cases, in which the anus exists in its natural condition, but opens into a *cul-de-sac*; the rectum being partially or wholly deficient.

3. Cases of imperforate anus, the rectum opening either into the vagina or urethra, or in some other abnormal direction.

Under one or other of these headings can nearly all the cases of this congenital defect be readily classed.

To Mr. Curling's exhaustive paper, published in the forty-third volume of the *Medico-Chirurgical Transactions*, we are unquestionably indebted for the bulk of our information upon this subject; and, as his description of these deformities with their assigned causes is so good, I may be pardoned for quoting him at some length.

It will be observed (says Mr. Curling) that the classification of these imperfections is founded on states which can generally be recognised during life. Unfortunately, the condition of the terminal portion of the intestinal canal, and its relations to the parts around, cannot be predicated with any certainty. In cases of imperforate anus, or of anus opening into a *cul-de-sac*, the intestinal canal may terminate in a blind pouch at the brim of the pelvis, the rectum being wholly wanting; or an imperfect rectum may form a short sac, descending to the floor of the pelvis, or so low as the neck of the bladder in the male, or the commencement of the vagina in the female.

It is known that the anal portion of the bowel is developed quite independently of the upper portion, and that the two afterwards approximate and unite, the diaphragm or septum disappearing by interstitial absorption. A failure in this process is the cause of the second form of congenital imperfection.

The cases of imperforate anus in which the rectum communicates with the urethra or vagina depend on the original existence of a *cloaca*, the malformation being due to an incomplete separation during fœtal life. These conditions are the result of an arrest of development at different stages. The blind pouch, in which the intestinal canal terminates, is sometimes connected with the anal integument, or with the anal *cul-de-sac*, by a cord prolonged from the bowel above. These cases are not, like the preceding, the result of a non-formation of the rectum, but are produced by an obliteration of the bowel, which was originally well-formed; the obliteration being a pathological change, due, probably, to inflammation which had existed during intra-uterine life.

The accuracy of these views is strongly confirmed by some cases quoted by Mr. Curling from M.M. Goyrand and Friedberg, where the muscular tissue of the intestine was clearly traced.

It is, therefore, with a view of determining the practice which should be employed in these different varieties of malformed anus and rectum, that I now purpose to enter upon the consideration of this subject.

Treatment of the First Class of Cases. In the first class of cases, in which the anus is closed or absent altogether, there would not be much difficulty in arriving at a correct conclusion as to the practice which should be followed, if it were possible for the surgeon to learn with accuracy by any examination the true position of the bowel, the terminal end of which is evidently malformed. It may be that the anus is the only part in fault, and that the bowel is natural as regards position, its orifice being alone closed. In these fortunate instances, there is not generally any great difficulty in determining the true condition of the parts; for a distinct bulging of the distended rectum will, in all probability, be readily felt on a careful manual examination; and in many cases the eye will at once detect the fact that a thin membrane alone obstructs the orifice. In such instances as these, there can be no doubt as to the practice which is to be pursued; a cautious incision over the spot in which the anus exists, or should exist, at once suggesting itself to the mind of the reflective surgeon. But this practice must be carried out with considerable care, and the finger should be the pilot. The incision, which should be tolerably free, must be carried backwards towards the sacrum, and not forwards towards the urethra or vagina, and under such circumstances, a free discharge of the retained meconium will in these simple cases be generally obtained.

In the more obscure class of cases of occluded anus, in which no such bulging of the bowel is to be seen or even felt, the practice which is to be carried out is not so certain, at any rate in its results; this uncertainty being due to the fact of its being perfectly impossible for the surgeon to form any opinion as to the true position of the malformed bowel. The rectum may be in its natural place, or it may not; it may terminate at the brim of the pelvis on the left side, or on the right, or it may not exist; and, under these diverse circumstances, the success of any exploratory operation in the perinæum cannot but be uncertain; although I believe it to be justifiable if carried out with care.

The plunging of a large trocar and canula into the pelvis with the vain hope of puncturing the distended intestine, is a practice which must be unhesitatingly condemned. It is unscientific on principle and unsuccessful in practice. It is injurious to surgery, as it is not based on a correct foundation; and injurious to our patients, as it is most fatal in its effects. Nevertheless, I believe that an exploratory operation in the perinæum, if conducted with care, is the best practice which can be primarily carried out.

An incision should be made in the same region as pointed out in the last class of cases—that is, in the position in which the anus should be, when naturally placed; and then, guided by the finger in the direction backwards and upwards, keeping the sacrum as the guide, the surgeon may fairly make an attempt to reach the bowel.

The tension of the distended intestines may be readily recognised, particularly when it has been once felt; and, having opened it, the operator should use all fair endeavours to draw down the intestine to the margin of the external opening, and to fasten it to the integument whenever it is found possible. But the chance of being able to succeed in doing this is not great, although the advantages which are to be acquired by success warrant the attempt.

The bowel, in the majority of cases, however, terminates at some distance from the surface. Mr. Curling states that "it is rarely found at a less distance than an inch from the perinæum;" and when we consider the shallowness of the child's pelvis, this depth is very great; the fact fairly indicating to us the probability that the rectum terminates in the majority of cases at the brim of the pelvis, and seldom dips down into it.

Out of twenty-six cases of this variety, which Mr. Curling has tabulated, the bowel was opened in fourteen; and in the remaining twelve, failure followed the attempt. Of the fourteen cases in which the bowel was opened, five only proved successful; whilst of the twelve cases in which the attempt failed, the bowel was subsequently opened either in the loin or groin in ten cases, and three of these died; the remaining two dying without anything further being done.

It must, however, be added that the great fatality of the instances in which the bowel was reached is to be explained by the rashness with which the exploratory operation was performed; a blind perforation of the parts with the trocar, with the vague hope of securing success, forbidding in many cases any prospect of recovery; the *post mortem* examination too often revealing the fact that the bladder, uterus, and peritoneum had been freely punctured, and that death was the result of the secondary peritoneal inflammation.

These figures prove to us the danger of the perinæal exploratory operation, and show that, unless it be undertaken with great care, more harm than good will probably ensue; nevertheless the propriety of making the attempt cannot be questioned, although it must be conducted with extreme caution.

Treatment of the Second Class of Cases. In the second or more complicated class of cases, in which the anus exists in its natural condition, but opens into a *cul-de-sac*, the rectum being partially or wholly deficient, the uncertainty of being able to form any opinion as to the true position of the lower bowel, renders any operative measure as hazardous as it has been shown to be in the former division; although it can be proved that, when the attempt to relieve has proved successful, a large proportion of the cases terminates well.

The upper portion of bowel may be in contact with the extremity or side of the *cul-de-sac*, or it may be some distance from it; but there certainly appears to be a stronger probability that the two extremities of the bowel will be found nearer in position in these cases, than in the former class in which no *cul-de-sac* exists.

The upper part of the bowel, in almost every instance, terminates somewhere at the pelvic brim; though it seems to be quite a matter of chance whether its terminal end lie to the right or left side of the pelvis, or on the sacral promontory; and, as the *cul-de-sac* is evidently an attempt of the lower part to meet the upper portion of the bowel, the distance between the two ends is not generally very great.

Under such circumstances, an exploratory operation at the extremity and posterior part of the *cul-de-sac* appears to be justifiable on principle; and, when conducted with judgment, it has also been practically found of value.

Mr. Curling gives us thirty-one examples of this class of cases. In twenty-seven an attempt was made to reach the bowel, and in sixteen the attempt was followed by success, and ten of these subsequently recovered. In eleven instances, failure followed the attempt; and, although other measures were carried out, all died. In four cases, colotomy was alone employed; three times in the groin with success, once in the loin with a fatal result.

In these cases, it would therefore also appear that an exploratory operation at the part is perfectly justifiable when conducted with caution; and that, if such should fail, colotomy is to be performed.

The exploratory operation must, however, be cautiously carried out; if a bulging at any spot can be detected, the attention of the surgeon should be directed to this part. An exploratory puncture may be made; a fine trocar and canula, I believe, being as good an instrument as any that can be employed.

It must be introduced carefully, and in the direction indicated, avoiding the anterior portion of the passage, and directing the point backwards and upwards. If any evidence can, in this way, be obtained of the bowel being perforated, the trocar may be taken out, and a grooved probe introduced in its stead, over which the canula can be readily removed. The opening may subsequently be freely enlarged by means of a bistoury, and the divided bowel brought down and stiched to the *cul-de-sac*, when found possible.

In the following case, the practice thus briefly sketched was carried out; and if a successful result forms a recommendation for the practice, it can in this instance be adduced.

CASE I. Edward B., aged 10 days, was brought to me at Guy's Hospital on January 15th, 1862, from Lower Norwood. He had not passed any motion since his birth, and had vomited freely. The child's general condition, however, appeared to be good.

On examining the part, a good anus was readily observed, and this led into a canal, which was about one inch long, ending in a *cul-de-sac*. The extremity of this was smooth, firm, and unyielding, and no evidence of any bulging of the bowel could be detected. An exploratory operation was, however, deemed advisable. I therefore introduced a fine exploring trocar and canula, no larger than a fine probe, into the upper and posterior part of the *cul-de-sac*, and, using considerable care in the exploration, came against an elastic, firm body, which was punctured, and the presence of fæces became tolerably clear; a grooved probe was then passed through the canula, and over it the canula was removed; by means of a curved bistoury, the opening was subsequently enlarged, and about an ounce of meconium came away. In half an hour afterwards, however, the bowels were freely relieved; the fæces being of a semi-solid consistence.

For domestic reasons, the mother was unable to stay at the hospital, and the child was therefore taken home. On the second day, the child appeared to be well in every respect; it had apparently been free from all abdominal pain; the motions had passed readily, and were of a healthy character; the abdomen was small, and all vomiting had ceased since the operation. The finger could be passed readily into the bowel, although a tight ring was at once observed. This digital dilatation was directed to be maintained daily; and Mr. Baker, the medical adviser of the family, kindly undertook to watch the case.

On March 7th, I had a letter from Mr. Baker, which contained a favourable report. Everything was going on well. The dilatation of the opening was persevered in, and no contraction had apparently taken place. The child's health was also good. Mr. Baker, however, added his fear that, unless constant dilatation was persevered in, contraction would follow.

The last report was on Dec. 10th; after which, the child having left the neighbourhood, Mr. Baker could give no intelligence.

In this instance, the success following the measures which were employed appears to me perfectly to justify the practice; and I do not think that a careful exploratory puncture by means of a fine trocar and canula can be looked upon as less scientific or valuable than a cautious exploratory incision with a bistoury. The evils of a reckless and incautious introduction of a trocar and canula cannot be too strongly condemned; but the same recklessness in the operator would attend the exploratory incision by means of a scalpel, and an equally bad result would probably follow.

To attain permanent success, however, in these cases, considerable attention is afterwards rigidly demanded. Repeated dilatation of the perforated bowel is absolutely essential to maintain its patency; for otherwise, like all artificial openings, its subsequent contraction will take place, with all the evils and fatal results of an ordinary stricture.

It will be gathered, however, from what has just been stated, that, in a large proportion of the cases of imperforate anus and obstructed rectum, a successful exploratory operation in the ano-perinæal region is not always to be achieved; for, in nearly half the recorded cases, or in 43 per cent., failure followed the attempt; and, as a consequence, the surgeon has in like instances to consider the important question as to the best means for affording any further prospect of relief.

Two operations suggest themselves for consideration: first, the opening of the bowel in the left loin, known as Callisen's or Amussat's operation; and, secondly, the one known as Littre's, or the opening of the bowel in the left groin.

The former operation, known as Callisen's, Amussat's, or the lumbar operation, is not one which offers many advantages. The irregularities in the position of the colon in the left loin, even in a healthy subject, become still greater in these cases of maldevelopment; and to the operating surgeon, they are, therefore, sources of considerable difficulty. The depth of the colon, when naturally placed in the young child; the probability of the kidney occupying the position of the colon; and the prospect of finding the bowel empty at the time of operation, are also additional arguments against its performance; but the strongest of all reasons remains to be repeated, and that is the bad success which has followed its adoption; for, out of seven cases in which it has been performed, five died. The operation is, therefore, one of uncertainty, difficulty, and danger; and these objections acquire greater force when it can be shown that Littre's, or the inguinal operation, is attended with more certainty, less difficulty, and a better success.

I need hardly repeat that Littre's operation consists in opening the sigmoid flexure in the left groin.

The integuments in this region are thin and readily divided; the bowel lies immediately beneath, and can without much difficulty be opened; and it can be hardly said that the inconveniences of an artificial anus in this region are greater than they are in the loin. In these respects, therefore, the inguinal operation has the preference; and the greater success which has followed the practice tends to support it still more. Out of fourteen instances in which it has been performed, nine recovered; and it will be remembered that two only out of seven recovered after the lumbar operation.

It can, therefore, I think, be fairly stated that the inguinal operation is the one which the weight of evidence tends to support; and that, in the cases in which a careful exploratory operation in the ano-perinæal region has failed to afford relief, a favourable result may with some confidence be looked for by pursuing M. Littre's plan.

Having, then, satisfactorily arrived at the conclusion that the inguinal or Littre's operation is the soundest practice to carry out, when relief cannot be secured by any operation in the anal region, we have next to determine whether the intestine should be opened on the left side, in which it was originally suggested, and has hitherto been practised; or whether it would not be a more certain and equally effectual operation to open the colon in the *right* groin. After carefully reviewing the chief facts connected with this subject, I believe that the best practice lies in the performance of this last suggestion; and my opinion is based upon the following reasons :—

Some portion of the large intestine will to a certainty be found in the right iliac region, whilst in the left this certainty does not exist; for it has been shown by Mr. Curling that in normal subjects, in which no deformity exists, in two instances out of twenty in which he practised this operation upon the dead child, the colon could not be found in its natural position; and, on subsequent dissection, it was found to be lying on the right side. In three out of the five subjects in which he also performed colotomy on the right side, the colon was readily reached, in one the cæcum appeared, and in the fifth case it is not stated that this intestine was misplaced.

In cases of maldevelopment, there is strong reason to believe that this irregular position of the descending colon exists in a larger proportion of cases; and that in such the terminal portion of the bowel will be found more frequently on the right side than in well developed and well formed children; or it may perhaps have assumed some median position. Under these circumstances the attempt to open it in the left groin would necessarily fail, and the life of the patient be probably sacrificed.

It appears, therefore, to me to be a more certain and equally scientific practice to open the colon in the right groin; for if the descending colon pass in that direction, it can be opened; and if otherwise, the cæcum can readily be secured; while, without doubt, under both circumstances, an equal relief will be afforded to the patient.

The difficulties of the operation in the right or in the left groin are the same; and there is a much greater chance of finding the large intestine on the right than on the left side. The relief which is to be expected by either operation is the same; for, I imagine, even if the cæcum or ascending colon be the part opened, there will be no difficulty in securing the emptying of the lower portion. At any rate, it will be no greater than is experienced in Callisen's or the lumbar operation.

In one case, which I now bring before you, this operation was carried out, and, although ultimate success was not secured, the following details will not supply any argument against the practice; but, on the other hand, I believe that they will materially tend to support the recommendation which I have just made.

CASE II. *Imperforate Rectum—Anus terminating in Cul-de-sac—Failure of Exploratory Operation in the Anal Region—Operation in the Right Groin—Death of the Patient.*—Eliza S., aged 17 days, was brought to me at Guy's Hospital, on July 31st, 1862. It was a well nourished, and apparently healthy child; and, with the exception of the absence of any action of the bowels, had been free from all symptoms of obstructed intestine till the ninth day, when some slight vomiting appeared. Since that date this symptom had returned at times, but was not in any way severe.

When seen, the abdomen was full, but not tense. The anus was perfect, and communicated with a *cul-de sac,* which was about one inch long. On a careful examination of the part, no trace of bulging of the bowel could be detected, and no indication of granulation at the upper extremity of the *cul-de-sac* could be observed. An exploratory operation, however, in the region of the deformity, was at once performed by means of the fine trocar and canula, but without success; consequently the opening of the bowel in the groin was at once decided upon. Chloroform was, therefore, administered to the child, and a careful abdominal examination made, when a curious coil of intestine became visible on the left side. There was no projection of the distended colon in the left iliac fossa, but rather a deficiency; and marked tympanitis existed upon that side. There was, however, a distinct prominence in the right iliac fossa, with palpable dulness, and from the left loin the intestine apparently passed across the abdomen, terminating in the centre beneath the umbilicus.

Having previously well thought over the operation which appeared to be the most applicable in cases of this description, and having determined that the opening of the colon in the right groin offered advantages which did not exist when the left was selected, I at once proceeded to carry out the practice; the case before me being apparently a good one for testing the worth of my views. I, therefore, made a vertical incision, at the distance of one inch from the anterior superior spinous process towards the umbilicus, and having divided the muscles and peritoneum, came down at once upon distended large intestine. Two thick ligatures were then passed through one margin of the integument, and carried completely through the intestine and integument on the other side; after which the bowel was opened, its thick, clayey contents at once escaping. The walls of the intestine were observed to be considerably thickened, and felt quite leathery, being evidently much hypertrophied. It must be added, also, that when the peritoneum was opened, a fimbriated growth, which turned out subsequently to be the Fallopian tube, made its appearance, and was, of course, at once returned.

The parents, contrary to all wishes, removed the child home; and some few hours afterwards, during a fit of coughing, the stitches gave way, and several feet of small intestine protruded. Medical advice was subsequently taken, but too late to afford relief, the child dying ten hours after the operation.

The following day, a *post mortem* examination was obtained, when the following appearances were observed.

About six inches of small intestine were extruded from the abdominal wound; these were dry, and of a dark colour. Upon opening the abdomen it was at once obvious that the descending colon had been opened by the operation; that some of the stitches had given way, and had allowed it to fall from the abdominal wall, small intestine passing in front of it, and protruding through the wound. The cæcum was situated to the outer and upper portion of the descending bowel; the transverse colon was placed in its natural site, and passed towards the left loin, to which it was connected by loose tissue and a distinct mesentery, proving that all attempts to open it by Callisen's operation would have completely failed. From this point it diverged from its natural path, and, instead of passing downwards over the left ileum to the pelvis, took a transverse direction obliquely across the abdomen over the sacral promontory, to terminate at the brim of the pelvis on its right side. The intestine terminated by a distinct pouch at the brim of the pelvis, and it was, therefore, a considerable distance from the anal *cul-de-sac*. The exploratory operation had done no harm, not having touched any of the pelvic viscera nor the peritoneum. The trocar had only passed into the cellular tissue at the posterior portion of the pelvis. It was then seen that it was the right Fallopian tube which had protruded from the wound during the operation, and was resting at its lower border.

It thus appeared that, by the operation in the right groin, the descending colon had been unexpectedly opened; and, had proper care been taken of the child, there is every reason to believe that a good result might have been secured.

This case must certainly be brought forward as a good one to illustrate the wisdom of selecting the right groin for the inguinal operation in preference to the left, and well supports the arguments in its favour which have been given in another portion of this lecture.

Should another opportunity offer I shall follow the same practice, believing it to be the best. I do not claim any originality for this suggestion which was made some time since by M. Huguier, at the Imperial Academy of Medicine, in 1858-59; but I believe that it has not been put into practice until the present occasion.

It would appear, therefore, that in both classes of cases to which our attention has been directed, a primary exploratory operation in the ano-perinæal region appears to be the best practice which can be followed. Great care and caution in the attempt are absolutely essential; and, should it fail, as past experience has proved that it will in nearly half the cases, the inguinal operation is to be preferred, the right groin being, I believe, the preferable seat.

In the subsequent treatment, constant dilatation is a necessary point of practice in order to prevent the recontraction of the part, which will otherwise to a certainty take place.

Treatment of the Third Class of Cases.—We will now pass on to the consideration of the third and last division of cases: in which the anus is imperforate, and the rectum terminates either in the urethra, vagina, or some other abnormal position.

It will scarcely be necessary to occupy your attention by describing all the vagaries of development which have been seen in practice at different times; for I believe that there is hardly a point in the median line of the perinæum, from the coccyx to the pubes, in which the rectum has not been seen to open; and that the vagina, from its orifice to its end, and the urethra, from its vesical origin to its close, have at different times been found to have received the contents of this malformed and ill-directed intestine.

The difficulties which are experienced in the treatment of these cases are peculiar to each; and, although every example must be treated on its individual merits, some general rules of practice may be laid down. It would appear, from Mr. Curling's researches, that it is into the urethra that the rectum most frequently finds an outlet; for out of forty-three cases which he collected, the bowel terminated in the urethra in twenty-six, in the vagina in eleven, and at some other part of the perinæum in six examples.

There is a practical point connected with this subject which claims our notice; and that is, in these cases of maldevelopment, the rectum, as a rule, is present, although misplaced. As a consequence, there is always a good hope that an operation in the seat of the natural anus will terminate with success.

When the urethra is the part implicated, the exploratory operation must be conducted in the anal region in precisely the same way as it has been recommended to be practised in the first class of cases which have been related, and in, which no outlet exists; and, judging from experience, the bowel will be reached in nearly two thirds of the cases operated on. The intestine must, however, be forcibly brought down, and firmly stitched to the integumental margin; for, if this point be disregarded, the opening will subsequently contract, and disappointment be the inevitable result. It is unfortunate, however, that, should success follow this effort for relief, nothing can be done in the way of closing the urethral orifice; if the anal opening be a free one and remains permanent, there will always be a solid hope of the abnormal passage in the urethra closing, but unless the former be kept open by surgical means, a double passage may remain for ever; the freedom of the flow of the fæces through the normal channel, being the only condition upon which the slightest hopes of the ultimate closure of the unnatural channel can be based.

When the vagina is the canal into which the rectum passes, a much better prospect of ultimate relief can be offered to the sufferer than in the last class of cases to which we have alluded. In these the outlet is often free, and, as a consequence, there is but little immediate danger to life arising from this maldevelopment; but the private and social disadvantages which the subjects of this deformity experience, render any operation which offers a probability of relief, not only justifiable but necessary.

In the majority of cases which have been operated upon, no difficulty has been experienced in opening the rectum in the anal region; for in eleven cases in which a record has been preserved, the gut was opened in all; and, in like instances, this same result can be readily secured by making a free section in the anal region upon a probe with a button end, bent so as to be made to project in the perinæum, after being introduced through the vaginal orifice.

The most difficult step of the operation, however, remains to be performed; and that is, to fix the walls of the intestine to the integument. This must be done by means of sutures; lateral incisions through the skin and tissue being made should any difficulty be experienced. If success follow this operation, an attempt to close the vaginal orifice may be made by a plastic operation; although the hope of obtaining a favourable result appears somewhat vague; for, judging from past experience, there is not one case on record in which it has been realised. In one instance, the ultimate closure was secured; but this closure was spontaneous, and it took place two months after the new one was established. In a second case now published the same result was obtained.

Two instances of this deformity have fallen under my care; and the following brief records of the cases may be read with interest.

CASE III.—Henrietta D., aged 8 months, was brought to me at Guy's Hospital on Nov. 8th, 1857, with an imperforate anus, and a small orifice communicating with the bowel at the mouth of the vagina. The child was well nourished, and apparently quite healthy.

Upon examination, this abnormal opening was found to be directed backwards, and to communicate directly with the bowel; and a bent probe, when introduced, was readily observed in the perinæum, in the position in which the natural anal outlet should have existed. A puckering, as of an anus, was also to be detected; and it appeared as if the orifice were closed by integument alone. An incision was therefore at once made in the site of the anus, the bent probe acting as a guide, and fæces at once escaped. In a week's time, when I next saw the child, the fæces had passed freely the right way, but little coming through the vaginal opening. There was a good control also over the action of the bowels; and in other respects the child appeared well.

She remained under my observation for three months; and, when she disappeared, the original vaginal orifice had completely closed, and the artificial and normal anus was firmly established. The control over the bowels' action was complete; and no signs of subsequent contraction had taken place. Indeed, the anus appeared to be as natural as if it had always existed.

This case cannot but be regarded as most satisfactory. It is true that an anus apparently existed, although it seemed to be closed by integument; and no great thickening of tissue had to be divided. The spontaneous closure of the abnormal opening, after the re-establishment of the natural outlet, with an absence of subsequent contraction of the anus, are points also of considerable interest.

Case IV. Ellen H., aged 8 months, was brought to me on May 14th, 1862, with the following congenital defect. The perinæum was altogether absent; there was not more than a quarter of an inch between the extremity of the coccyx and the median fissure, which fissure extended forward to the pubes, and was bounded by labia. In this fissure were the orifices of three canals. Each canal was divided by a membranous septum, and their extremities were most distinct. The posterior canal was small, not being larger than a swan-quill; it was also the most prominent. The vaginal tube was large and very patulous, its anterior wall apparently falling forwards; the orifice of this canal being evidently less advanced than the anal; and above this passage, and more deeply placed,

D

was the opening of the urethral canal. In other respects the child appeared perfectly sound and well developed. There was no control, however, over the bowels' action; fæces were always passing with perfect facility, and there was evidently a complete absence of sphincter.

In this case nothing could apparently be done. The outlet of the bowel was sufficiently free to allow of its complete evacuation; and there was nothing to lead one to hope that the least benefit could be obtained by an operation. The peculiarity of the case, indeed, appeared to consist in the complete absence of the superficial perinæal muscles, and the distinct presence of the three canals—the rectal, the vaginal, and the urethral.

As a general summary of the subject, the following conclusions may be given.

1. In all cases (with some rare exceptions), whether of imperforate anus, obstructed rectum, or misplaced anus, an exploratory operation in the normal anal position is perfectly justifiable; and it may be attempted with the fair hope that, in nearly half of such cases, primary success will be secured.

2. Such exploratory operations, however, to be successful, must be conducted with great caution; and the exploratory puncture or incision is to be made upwards and backwards towards the sacrum.

3. If these measures fail, or if from some peculiarity in the case they appear useless and unjustifiable, the intestine is to be opened in the

groin; the lumbar or Callisen's operation being quite inapplicable.

4. In the inguinal operation, the right groin appears to possess advantages over the left; as the intestine is found with more certainty, and the benefits to be expected from the operation are equally great.

5. The treatment of these cases does not terminate with the success of the primary operation; for constant dilatation of the artificial anus is a necessity to preserve life.

END OF FIRST LECTURE.

LECTURE II.

Mr. President and Gentlemen,—In my former lecture, I briefly brought before your notice the chief points of difference between the physiological and pathological processes as witnessed in early and in adult life; and then passed on to the consideration of some of the congenital defects which are to be observed in the oral and anal regions. The abnormal developments of the nervous, genito-urinary, and osseous systems, would next claim our attention; and had time allowed, I should have been glad to have brought these subjects under your notice. In the limits of three lectures, however, it is absolutely impossible to include every point which may be worthy of attention in the surgical diseases of children; and I am, therefore, compelled to pass over these, with many others of equal interest.

I must confess that it is with much regret that I am obliged to dismiss from our consideration the interesting subject of *Maldevelopment of the Head and Spine*, as exhibited in cases of meningocele, encephalocele, and spina bifida; but, as we are able to refer to the labours of such an accurate and scientific investigator as Mr. Prescott Hewett, who quite exhausted these subjects in his admirable lectures delivered at the College of Surgeons some few years since, these regrets are materially modified; for I feel that I do not possess much new material to add to our stock of knowledge, or any new views which might throw additional light upon the nature and treatment of this important class of cases.

Hermaphrodism is another subject which invites investigation; but, as it would require the three lectures to do it justice, it must be set aside, with the malformations of the penis and urethra.

I must quote, however, a singular case of arrest of development of the osseous system, which manifested itself by the absence of certain bones; and it is to be remarked that the left side of the body was the part affected.

Absence of Fibula, Os Calcis, Cuboid, and Three outer Metatarsal Bones, with the Toes. Robert F., aged 15 months, was brought to me at Guy's Hospital on Feb. 21, 1859, with the following deformity. The child was apparently quite healthy and in good general condition. The left leg and foot were, however, much wasted, and had been so from birth; but the extremity was of its normal length.

A careful examination showed the following conditions. The thigh appeared to be quite natural; but in the leg there was no evidence of the presence of any fibula; neither its head nor external malleolus could be detected, and there was little soft tissue to obscure the examination. The foot was curiously ill-formed; it had but two toes, corresponding to the great and second toes. The four outer were absent, with their metatarsal bones, the cuboid, and os calcis; the astragalus, scaphoid, and cuneiform bones, with the metatarsal and phalangeal bones belonging to the two inner toes, being alone present. It was difficult to make out what muscles existed. The calf of the leg was very flabby, and the tendo Achillis was apparently absent. The toes could be moved by muscles and were well developed. The anus was situated somewhat towards the left side, and the left testis had not descended, but its distinct scrotal pouch was very visible.

Orthopædic surgery is another subject which might claim a passing attention, as a large number of cases come under our care during each twelve months; but it has been so ably written upon by many specialists, that I feel it would be presumption on my part to take up your time by its consideration. I will, however, make one observation upon the treatment of these cases, which experience has taught me; and that is, that by far the larger proportion of cases may be successfully treated by mechanical means alone, and without operation—that is, without the division of the contracted tendons.

The rule which I apply to these cases is as follows: that if the foot can be restored with but little force to its normal position, a cure will, in all probability, be secured by simple means—that is, by extension; and that the operation of tenotomy may be set aside; but if much force is called for to bring about this result, the contracted tendons should be divided.

The mechanical means which I have been in the habit of employing are very simple; a good firm linen strapping being all that is required by way of material. The method of its application is also very simple, and may be thus described. If the case be one of varus or equino-varus, a piece of strapping about one inch wide, broad enough to cover the body of the foot, and about nine inches long, should be selected, and fastened by at least one circular turn round the foot, leaving the end on its outer side; the foot is then to be brought to the required position, and fixed there by forcibly bringing the extremity of the plaster to the upper and outer part of the leg, using it as a side splint. This upper extremity is then to be firmly held in position by an assistant, whilst a second piece is applied round the foot and ankle as a figure of 8. By this turn, the external upright plaster splint is firmly compressed towards the leg, and, thus drawing upon the ankle, turns the deformed foot firmly outwards, and fixes it in the required position. A third or circular piece may then be applied round the leg to fix the upper portion of the vertical plaster splint, and the thing will be complete.

If these strappings be well applied, it will not be

necessary to readjust them for at least three or four days; by such means a large proportion of cases will be readily and permanently cured.

The principle of its application rests in the extending force which is employed by the external or vertical portion: the second or figure-of-8 binding this vertical splint firmly to the leg, increasing its extending force and making it permanent.

I have carried out this practice extensively at Guy's Hospital with great success; 'and, as the material is inexpensive, and its application is easy when once learnt, I have thought it worth bringing before your notice.

It is hardly necessary to quote examples in which this treatment has proved of value. My note-books contain many such; but the fact of its proved usefulness amply suffices for my present purpose.

In cases of talipes equinus and valgus these plaster splints are equally applicable; the surgeon having only to modify the application of the principle to each form.

With these short observations upon as many of the special surgical affections of early life as the period allotted to these lectures will allow me to make, I must now pass on to the consideration of other portions of my subject, and point out the differences which are found in practice between the injuries and diseases common to adult and infant life, and the modifications of their treatment which are consequently required. I propose to bring this subject before your notice by taking the different systems in rotation, commencing with the injuries and diseases of the nervous system.

THE DIFFERENCE BETWEEN THE DISEASES OF THE NERVOUS SYSTEM IN THE CHILD AND IN THE ADULT.

The marked anatomical differences which are to be observed between the crania of the adult and infant during the first few years of extra-uterine life, and the equally important and palpable differences between the nervous centres of the human species at these two periods of existence, are of themselves amply sufficient to lead any reflective mind to the conclusion that there must be some important difference between the diseases of such structures at these opposite ages; and the results of experience prove the positive correctness of such an inference.

The anatomical purposes which are embodied in the development of the *adult* cranium are, without question, essentially *conservative*. They are perfected slowly, and with marvellous precision, and are adapted with admirable accuracy for the security and preservation from external injury of the all-important and wondrous structure, man's brain.

The anatomical purposes of the *child's* cranium are also conservative, but conservative only to a degree; for this high purpose is evidently completely subservient to the higher and more important conditions which appertain to the growth and development of the cerebral structure. In the adult cranium the anatomical purposes are solely adapted for the protection of the *perfect* brain. In the infant or child cranium the anatomical purposes are mainly adapted for the growth and development of the *perfecting* brain.

Respecting the structure of the brain, the firm and consistent nature of the adult organ stands also in bold contrast to the soft and pulpy nature of the child's; and the established functions which appertain to the perfect organ form an equal contrast to the apparently uncertain and growing functions of the imperfect.

This activity in the growth and development of the child's nervous centres is associated also with an increase of sensibility, and a degree of sensitiveness to centric or eccentric impressions, which are not to be met with in the adult brain; and it is due to this fact that we are enabled to explain the differences which are to be observed between the symptoms following injuries to the head in childhood and manhood.

It must not, however, be for one moment thought that this excessive irritability of the brain is manifested after every slight injury; for experience tells us that this is not the case, and that children can bear blows and falls upon the cranium with perfect impunity; the anatomical disposition of the bony centres being such that they invariably receive the first impression of the fall, and their membranous margins or fontanelles, taking up the impulse, retard and stop the force of the onward vibrations. The brain is not, therefore, as a rule, so materially shaken by a slight injury as, in surgical language, to become the subject of a concussion. But if this condition be experienced, the symptoms denoting cerebral irritation are generally severe and most marked. When once originated also, the disposition to excessive reaction or to inflammatory complications is very great; and in the following case the truth of these remarks is well illustrated.

CASE. Arthur B., aged 3, was admitted into Guy's Hospital, in July, 1860, under the care of Mr. Birkett; but the treatment of the case fell into my hands. Shortly before admission, the child had fallen from a window eight or ten feet high, pitching upon his head. He was taken up hastily, and brought to Guy's, when he became conscious, presented no external signs of injury, and no symptom of cerebral mischief could be detected. He was placed in bed, and ordered to be kept quiet, as a precautionary measure; and everything went on well till the third day, when fever appeared, accompanied with vomiting, unconsciousness, and the peculiar shrill scream so characteristic of cerebral irritation. The head was tilted back upon the spine, and the pupils were dilated. A spirit lotion was prescribed, for the purpose of keeping the head cool; and a mild mercurial of grey powder, in three grain doses, with half a grain of Dover's powder, given every six hours; and in three days these symptoms had subsided. From that date everything went on well, and convalescence was established.

Many cases somewhat similar to the above might be selected from my notebook, although, perhaps, none could be more typical of the class of injuries. The necessity of preserving absolute rest in even the slightest injury to the skull is a rule which it is hardly necessary for me now to dwell upon, although it is difficult to make the inexperienced understand this necessity when no symptoms of brain-mischief exist. In children this rule is as absolute (if not more so) as it is in adult life, although there may be some difficulty in carrying it out. For it must be remembered that, as there is in childhood a greater activity of the vital functions in the cerebral centres for the purposes of their growth and development, so is there a greater tendency to cerebral irritation and inflammation; and in proportion as the organ is more delicate in its structure than it is in adult life, so is the result of morbid action more important, as well as more rapid and destructive. As a consequence, therefore, injuries to the cranium in young life are followed by symptoms of concussion more frequently than they are in adult life; and, when once manifested, these are liable to become more severe. The child-brain, from its softness and want of consistence, is also more liable than the adult brain to become the subject of ecchymosis from slight causes; and it is due to this fact that cerebral symptoms from secondary inflammatory action are so frequently observed.

It can be no matter of surprise, therefore, that a fatal result is by no means uncommon after a slight cerebral injury during child-life.

The delicacy of the brain and its wonderful irritability are well shown by the fact that every symptom of severe disease of its structure may make its appearance, and death follow, yet no pathological indication may be detected upon subsequent examination.

It is hardly consistent with received opinions or our own belief to acknowledge that altered function suffices by itself to destroy life; and yet, if the absence of positive evidence of cerebral disease in cases in which every symptom of cerebral affection not only existed, but proved fatal, tends to prove the truth of such an opinion, it must be here adduced.

For a child may, without doubt, die off-hand from simple concussion, and no evidence of organic change be subsequently seen on *post mortem* examination: and a child may also die during a convulsion produced by a local injury or some reflected eccentric cause, with the same subsequent absence of local pathological conditions.

Again: as another proof of this excessive irritability of the nervous centres, so predominant in early life, the effect of injuries to the extremities or integument may be brought forward; for it is, I think, certainly true that tetanus and convulsive diseases are by no means unusual complications of these injuries; and it is this fact, if such it may be called, which must tend to modify our practice in the treatment of these cases. It has fallen to my lot to see several fatal cases of tetanus following upon laceration of the integuments and compound fractures in young subjects, in which conservative surgery was carried out, when perhaps a more energetic practice might have saved life, although with the loss of a limb; and I have it upon my conscience that I should in some few cases have been led into such an error, and have thus lost my patients.

It is almost an impossibility to bring forward positive proof of the accuracy of these views; but it is certainly within the limits of experience when I assert that it is a rare thing for tetanus to appear in an adult after a lacerated wound or compound fracture, and these injuries are not uncommon; but in young life, although such accidents are rare, death from tetanus is by no means unfrequent.

It is, therefore, with some confidence that I now venture to assert my belief that tetanus is a more frequent result of a lacerated wound, or of a compound fracture, or of any other integumental injury, when these accidents occur in young subjects, than when they take place in the adult; and that, therefore, so-called conservative surgery in such cases must be more cautiously carried out in children than in adult life. As additional evidence of this opinion, it may be stated that, in Mr. Poland's able communication on Tetanus in the *Guy's Hospital Reports* for 1857, out of seventy-two cases which he brings forward, fourteen were in children under fifteen years of age; this being one-fifth of the whole number of cases.

The following two cases will perhaps tend to illustrate the truth of this record.

CASE I. William E., aged 5 years, was admitted into Guy's Hospital, under my care, on September 8th, 1860, for a severe lacerated wound of the left elbow, exposing the joint, produced by the passage of a cart-wheel over the part. The vessels and nerves were, however, apparently uninjured. As the local injury was not believed to be beyond recovery, a splint was applied, and iced-water dressing employed. Everything went on well till the sixth day, when tetanus appeared, followed by death in ten hours.

CASE II. Eliza W., aged 5, was admitted under my care at Guy's, on April 8th, 1861, with an extensive lacerated wound of the left leg and fracture of the tibia and fibula in their upper third, from the passage of a cart-wheel over the part. On the ninth day, tetanus came on; amputation was performed without benefit, death taking place in sixty hours.

It is quite unnecessary for me to enlarge upon the treatment of the injuries to the skull and its contents in early life; for the same principles which are applicable to the adult are equally so to the child; my object being to dwell upon the peculiarities or differences which are to be found in practice, and not on the points in which they agree.

The surgical diseases of the cranium need not also occupy our attention for long, as they differ but little in their nature at the two ages. Inflammations, however, of the cranial bones in children are not common; and the following case will therefore be read with interest.

CASE. Martha J., aged 15 months, was brought to me at Guy's Hospital on March 7th, 1860. She had been ill two months, and, when seen, was suffering from extensive necrosis of the left half of the frontal bone, consequent on a wound received from a nail. There were no head symptoms, nor any indications of general mischief. The bone was readily removed by means of forceps; and the dura mater was exposed, covered with granulation. Everything went on well, and recovery took place quickly.

In this case, the rapid progress of inflammatory disease of the bone was well illustrated; and its death and subsequent removal were accomplished without giving rise to any cerebral symptoms. In this point of view alone the case is, therefore, worthy of record. In adult life, it is not a common thing to find necrosis of the whole thickness of a cranial bone; and when we do, it is not a common thing to find it progress without giving rise to cerebral symptoms. Syphilitic necrosis in adult life is the commonest form; and it is too true that, in these cases, epileptic seizures as a result are not unfrequent; these seizures being clearly due to an extension of the inflammatory mischief to the dura mater and membranes beneath, and subsequent cerebral disease.

In childhood, these cases are unknown; the disease of the cranial bones being generally due to some accident similar to that noted in the above case.

DIFFERENCES BETWEEN THE SURGICAL AFFECTIONS OF THE RESPIRATORY SYSTEM IN THE CHILD AND ADULT.

We will now pass on to consider in what way the surgical affections of the respiratory organs in children differ from those which are to be found in the adult; and if there are none which may be described as special to this early period of life, there are many which require surgical assistance, and which are, as a rule, observed only in young subjects.

Amongst the most important of these affections

is that *œdema of the larynx* which follows the swallowing of boiling liquid; and the symptoms which immediately result from such an accident are mostly severe and dangerous to life. The pathological condition which results from the contact of the fluid is of a simple character, a severe blistering of the part being the prominent point; the œdema and vesication of the irritated mucous membrane readily closing the narrow rima.

The principles of treatment in such cases are not complicated, and they are the same as would be carried out under any other condition. But delay in their application is not to be allowed; for suffocation, either from the mechanical obstruction occasioned by the enlargement of the parts, or from spasm of the glottis, is most likely to ensue. If the symptoms, however, are not severe, and if, on careful watching, they do not rapidly increase, there is strong hope that the mischief is but partial, and that they will soon subside; for this blistering of a delicate membrane is a rapid process; and if it do not take place within a few hours after the accident, it will rarely appear. Nevertheless, the child must be watched with care, so that the evils of a sudden spasm of the glottis may be averted, and the earliest indications of obstruction successfully treated.

Dr. Bevan and others have spoken in high terms of the benefits of antimony in this affection; and the weight of evidence which they adduce in its favour must be considered strong, and, therefore, highly recommendatory. It is to be given in small and frequently repeated doses. I have given it in cases of the milder description with some success, when no immediate danger to life manifested itself, and when delay was, therefore, not so dangerous; but I have never yet been induced to trust to it when there was an evident fear of a momentary suffocation, and therefore an imminent danger to life. In such severe instances, some immediate means to prevent suffocation are loudly called for; and of these, tracheotomy has been the most favoured.

I possess the records of nine instances in which this practice was carried out, and in five recovery took place; in the remaining four, death followed from bronchopneumonia. I am aware that this success is greater than that usually recorded; for out of twenty-five cases recorded by Dr. Jameson in the *Dublin Quarterly Journal of Medical Science* for February, 1860, and the *Medical Times and Gazette* for October, 1859, but six recovered.

It is not an easy question to decide how much of this pneumonia is to be attributed to tracheotomy, or to the accident; but the operation is almost a necessity to prolong life, and must therefore be performed. I must confess, however, that I have derived much satisfaction from a suggestion which has been made within the last few years, but which, I believe, has not been carried into practice; and that is, to puncture or scarify the œdematous mucous membrane. If the pathology of the affection be such as I have indicated—merely a blistering of the mucous surface—there are great hopes that this practice may prove of value, and I shall certainly be disposed to give it a trial on a befitting occasion. It has this advantage, that on principle it is likely to succeed; and, if the practice fail, no possible harm can accrue, as other means can be at once employed.

The subject of *foreign bodies in the air-passages* is the next to which I propose to draw your attention; for, although cases illustrating the evil effects of such an accident are occasionally met with in adults, by far the larger proportion of these instances are undoubtedly found during early life.

It is scarcely necessary to dwell upon the symptoms by which this accident is to be diagnosed. That obstruction to the respiratory process exists, that it came on suddenly, and is so aggravated at intervals as to endanger life, are symptoms sufficiently characteristic to indicate the nature of the case. The history will also, as a rule, prove of material assistance. But any one who has witnessed such an accident will find no difficulty in readily recognising a second.

There is but one principle of treatment which demands attention, and that is, to procure, as speedily as possible, the expulsion of the foreign body. Without some surgical measure, it is an extremely rare thing to witness this event, when the foreign body has passed the rima, and entered the trachea; and, although I have read of such cases, it has never fallen to my lot to witness one.

An operation, therefore, is almost always called for; and that of tracheotomy is the only feasible one. I possess the records of eight examples in which this operation was performed. In five of these the foreign body was immediately ejected upon opening the trachea; and four of these recovered, death occurring in the fatal case from bronchopneumonia. In the remaining three cases, the foreign body was immoveable, and produced death. In two it was subsequently found to be firmly impacted within the rima; and in the third it was impacted in the right bronchus.

It thus appears that in a large proportion of the cases in which a foreign body has passed into the trachea, its removal can be successfully obtained by the operation of tracheotomy. Indeed, as long as any evidence exists that it is still free and moveable, and is not impacted either in the bronchi or larynx, a successful result to operative interference may with some confidence be predicated.

If the foreign body be impacted within the bronchus, there is little hope of its extraction The use of forceps and other instruments which have from time to time been suggested by surgeons for this end, cannot be recommended; for it is almost an impossibility to seize the foreign body when impacted within the tube, and any attempt will, as a rule, only tend to fix it more securely in its position. If air can pass into the lung, there must always exist a strong expulsive force, which powerfully tends to procure the removal of the body; and if it be fixed in too firmly to allow of such a result taking place, any attempts from without will be obviously futile. In exceptional cases, however, this practice may be of use. In all cases, therefore, of foreign body in the air-passages, tracheotomy should be performed, and delay in its execution must be unhesitatingly condemned, as no possible good can follow such a line of practice; the surgeon must not look for the natural expulsion of the foreign body through the rima, for so long as such remains within the trachea, the laryngeal orifice is pretty sure to be perfectly closed, although recorded experience favours us with instances in which the foreign body was spontaneously expelled from the trachea without an operation. Chloroform should be administered preparatory to the operation, as it may be given with perfect safety; for it not only tends considerably to prevent the spasm of the laryngeal muscles by allaying their irritability, but enables the surgeon to perform his difficult duties with calmness and precision.

A free section of the trachea should always be made in order to allow of a good passage for the expulsion of the foreign body; and, in the majority of cases, this foreign body will be expelled as soon as the opening in the trachea has been accomplished. If any difficulty be experienced in obtaining this result the child's body may be tilted upwards and the head downwards, to favour the gravitation of the substance, and a good shake or pat upon the back will at times dislodge it from its position, and consequently assist in securing its expulsion. If these means fail in carrying out the object for which they were undertaken, the child must be left alone; the opening in the trachea should, however, be made large, or even valvular, and means adopted to preserve its patency, so that, should the foreign body by any chance be subsequently dislodged from its position, its expulsion may be secured.

The larynx itself should be always most carefully examined, not only before, but after the operation; for it is not an uncommon thing to find the foreign body firmly impacted within the rima, and consequently in a position from which its removal cannot be looked upon as being very difficult. In two of the cases which I have recorded, out of the three which proved fatal after a futile attempt at the removal of the foreign body by tracheotomy had been made, I have already shown that the foreign body was subsequently found to be firmly impacted within the rima; and it is fair to believe that if the foreign body had been previously removed from this position, a very different result would, in both cases, have ensued.

This laryngeal exploration must, however, be conducted with considerable care. The passage of a probe or fine catheter, or any other small body, through the rima from below upwards, is positively useless, as experience has proved that these means may be carefully employed, and yet the surgeon will fail in finding any obstruction. The exploratory instrument must be a large one; indeed, it should be nearly as large as can be admitted through the rima: so that if the foreign body lodge at this spot, it will necessarily be pushed upwards into the pharynx by its forcible passage. A large elastic catheter is unquestionably the best instrument for this purpose; and if this be employed in such cases, it may be unhesitatingly asserted that a more uniformly good success will in future have to be recorded. This practice would probably have added two more successful cases to the five which I have already mentioned, making seven successful instances of the removal of the body, out of the eight in which it had taken place.

Within the last few months I was present at the examination of a child who had died from the impaction of a bone within the rima, a fact which had not been detected after the operation of tracheotomy, although a careful examination of the larynx was made with a fine instrument. It was impossible to see the specimen without a painful feeling of regret that a larger instrument had not been employed in the exploration, as the body was at once dislodged after death by the introduction of a catheter up the passage.

In all cases, therefore, of foreign body in the air-passages, in which an expulsion cannot be obtained by the operation of tracheotomy, let the larynx be carefully examined by means of a large catheter, and its extremity passed from below upwards through the rima, the finger at the same time making a careful examination from above; for by these means, and these only, will a foreign body be removed from the rima glottidis.

When the body is removed, convalescence may fairly be anticipated. An attack of bronchopneumonia will at times interfere with recovery, or even produce death; and, although this complication is one to which the attention and fears of the surgeon may be directed, it is not one which frequently follows, or which should produce other feelings than those of anxiety.

Foreign Bodies in the Nostril are common occurrences in children. It is a subject, however, which need not occupy our attention. They may readily be removed by means of a bent probe or forceps, and if of a soft nature, a good syringing answers every purpose.

I have known a plum-stone to remain within the nostril for eight months, and give rise to symptoms of ozæna, for which the child was brought under my observation. A careful examination of the nostril in such cases will, however, invariably reveal its true nature.

The surgical affections of the upper part of the respiratory organs, having received as much of our attention as we can afford to give, I propose now briefly to consider if there be any points of difference to be noted between the surgical affections of the adult and child, and I am disposed to believe that the most important difference is shown by the fact that in young life an injury to the thorax may be followed by laceration of the lung-tissue, without any associated fracture or displacement of the ribs. This fact is unquestionably to be explained by the greater elasticity of the thoracic walls in the child than in the adult, and the greater tendency to laceration of the lung-structure. The following case will, perhaps, best illustrate its truth.

CASE.—A boy, aged 7 years, when playing in the road, was knocked down by the shaft of a cart; the wheel caught him on the left side of the lower part of his abdomen, and turned him round, stopping when just about to pass over the thorax. Intense dyspnœa and severe hæmoptysis immediately resulted, and he was brought to Guy's. He was admitted under the care of Mr. Birkett; and, as I happened to be at the hospital at that time, I saw him, and noted the following facts. He was in bed, lying on his right side, half turned over on his abdomen, with his hips drawn up and flexed. There was intense dyspnœa and cough, accompanied with hæmoptysis. He was quite collapsed, and nearly unconscious; and no indication of fractured ribs could be detected. He never rallied, but died two hours after the accident.

At the *post mortem* examination, the only external sign of injury was a bruise on the left side of the back. There was no fracture of the ribs, nor any external indication of injury to the thorax. The right chest was filled with air and some ounces of blood, which had evidently proceeded from a laceration, about three inches long, of the lower edge of the middle lobe of the right lung. The lung was partially collapsed. In the abdominal cavity were a few ounces of blood, from a laceration of the upper edge of the liver. There was also effused blood about the left kidney, from laceration of the suprarenal capsules.

REMARKS. In this case, I think, there can be little doubt that the laceration of the lung was due to the injury; and, as it was not complicated with any fracture of the ribs, it must have been produced by a yielding of the bony and cartilaginous coverings, and a dragging of the lung from its fixed attachments. This is a point of considerable interest, and, not being common, is worthy of record. It is mentioned in this place as a principal point of difference between the thoracic injuries of young and adult life, and is the only one to which I will now draw attention.

Fractures of the ribs in children, like the same injury in adult life, are to be treated by strapping, and not by bandages; bands from one to one-and-a-half inch wide being fixed from the spine to the sternum round the injured side, including three ribs above and three below the seat of fracture. It must be remarked, however, that in young people fractured ribs are very rare; their natural elasticity, doubtless, being quite sufficient to explain the fact.

DIFFERENCES BETWEEN THE SURGICAL AFFECTIONS OF THE CIRCULATORY SYSTEM IN THE CHILD AND ADULT.

THE differences between the surgical affections of the circulatory system of the child and adult are the next to which our attention must be directed; and the subject of *nævus* perhaps claims our first notice. That it is an affection of young life every one will admit; and I imagine few will deny that it is a disease of the vascular system. Whether it be a new growth altogether, allied to the erectile tumours, or a simple increase or multiplication of normal capillary tissue, may be open to dispute; but I am disposed to explain the appearance of nævi by this last suggestion, and to suppose them produced by a multiplication or growth of capillary tissues; the formative and developmental forces which are so active in young life, taking a wrong path, and directing their power simply to the repetition and growth of capillary tissue, instead of the healthy development of the natural body. This misdirected force may also continue its action for some months; but it has always a tendency to cease, for it is only during the first few months of life that nævi show any rapid signs of increase.

The subsequent life of nævi tends also to bear out the truth of this suggestion. In many cases they continue to grow for months with varied degrees of rapidity, and then suddenly cease; remaining either in a stationary condition, or, as is more frequently the case, undergoing a degenerating process. In some instances, this degeneration of the growth commences *in utero;* when, at birth, decaying or so-called ulcerating nævi will be observed; one and all of these nævi, at different times of their existence, ceasing to increase; and, one and all, at different epochs of their life, beginning to degenerate; the growing power after a time ceasing to manifest itself, as the developmental power had previously done. Obedient also to the physiological law which is in force during all ages—more particularly during young life—and which shows itself by the tendency that all parts of the body possess to return to their normal and healthy condition, so these nævi degenerate and waste, and cease to trouble except by their deformity.

Nævus is, therefore, essentially a disease, or rather a growth, of young life, making its appearance either at birth or a few weeks subsequently, and is found only when growth and development are the most active; and it is to be explained by an excess in the growth of capillary tissue, and a want of natural development. The treatment of these cases will next claim our attention.

Treatment of Nævi. It may with some truth be stated that, unless a nævus be so situated as to be an eye-sore or an inconvenience, and unless it show positive evidence of its tendency to rapid increase, there is no necessity for operative interference; it will, to a certainty, after a time, cease to grow, and will also to a certainty begin to degenerate or waste; and under such circumstances, in a large proportion of cases, it is hardly necessary to interfere. Should, however, the nævus be so situated that it is or will be either an inconvenience or a deformity; or should it grow so rapidly as to threaten to become either, something must be done; and this something is to be determined by the nature of the nævus and of the tissue in which it is placed. It may involve either the skin alone, or the cellular tissue alone; or it may involve both.

In the first class, in which the skin alone is involved, the nævus may readily be destroyed by means of external applications, such as caustics, nitric or sulphuric acids, potassa fusa, chloride of zinc, tartarised antimony, etc. The caustic must be used freely, and without fear; and one application should be amply sufficient. In this *simple cutaneous* form, external applications are to be employed, and in it alone; in all other forms they are absolutely useless.

In the purely *subcutaneous* form, removal by the subcutaneous ligature is doubtless the best practice. The whole nævus must be freely surrounded, and completely strangulated; any surgical knot which will carry out this object answering the purpose.

In the *mixed* variety, excision of the growth, when it can be performed, is to be recommended; but if this plan be inapplicable, a strong ligature, including the nævus and integument, should be employed. Excision is the best practice when the nævus is pendulous, or when it can be previously isolated from the parts beneath; a clamp having been applied along the line of section, to prevent hæmorrhage. In several instances, I have excised the nævus; having previously inserted some pins beneath the growth, and cut upon them, keeping the pins as points round which a ligature might be applied, and by which the edges might be subsequently brought together. The operation of excision is doubtless one of risk, and much care must be taken that the nævus itself be not touched; for this purpose, the line of incision should be far from its margin. It is only in select cases, however, that excision is to be performed. If the thickness of the lips be the part involved, such a practice is the best; and in two cases, one of the upper and one of the lower lip, I have carried out this practice with perfect success; the edges of the V incision having been brought together, after its removal, by a fine suture. In the majority of cases, however, of the mixed variety, the ligature should be preferred.

It occasionally happens that in some instances none of these plans of treatment appear suitable. The nævus may be too diffused, or too close to some important part; and, under such circumstances, some other plan of treatment is to be employed.

In the subcutaneous or in the mixed variety, the treatment by injection of the perchloride of iron is a good one, although it is by no means generally successful. When it answers, however, it succeeds well; and it is, therefore, worthy of a fair trial.

The treatment of a diffused nævus by means of setons is another plan of treatment, and it is one to which I must confess to be very partial. It has proved of great value in my hands; and I give, therefore, a preference to it over injection. I could, if it were necessary, quote numerous instances in which success has followed the practice; but the fact of its success is all that requires to be established.

If the nævus be large, several thick setons, passed through the centre of the growth, may be inserted; but, as a rule, one or two amply suffice. By these means sufficient inflammatory action is produced to close temporarily the capillary tissues, and, by subsequent contraction, to effect a permanent cure.

DIFFERENCES BETWEEN THE SURGICAL AFFECTIONS OF THE DIGESTIVE SYSTEM IN THE CHILD AND IN THE ADULT.

In the digestive system there are no special surgical affections which belong to early life, excluding the congenital deformity of hare-lip, which has already received attention; but I possess the notes of some few interesting cases which appear worthy of record.

Salivary Fistula is a rare affection both in early and in adult life; and it is somewhat difficult to understand how it can be produced, unless an injury should have been the cause.

CASE. In 1858, a boy, aged 7, came under my care, with a fistula in his left cheek, which had been discharging a clear fluid for one month. A swelling had previously existed at the spot for one year, and this had gradually increased until it had acquired the size of half a walnut. The mouth was dry upon that side; and the boy stated that, when eating, the discharge from the cheek was very profuse. This fact was readily tested, and proved to be correct, and several drachms of saliva were quickly obtained. Upon exploring the cyst with a fine probe, no opening into the mouth could be detected. I therefore passed a seton through the cheek, and left it in for two weeks. By that time the saliva had found its way through the artificial opening; and, as a consequence, the seton was removed. The external wound rapidly closed, and convalescence was secured.

Ranula, or rather *sublingual cyst*, is not a more common affection in early than it is in adult life; but when it appears in children, its increase is certainly more rapid. But this is only consistent with the fact that all secretion is more active at that early period of life. Ranula is now well known, from the investigations of recent pathologists, to be simply an obstruction to the ducts of the mucous glands, and is not in any way connected with the salivary. In children, the secretion of mucus is always active and profuse; and on that account cases of ranula are generally rapid in their appearance. The most effectual treatment is to be obtained by the passage of a seton through the tumour, a double thick silk one being the best. It is seldom necessary to leave it in more than one week, sufficient inflammation having been excited during that period to cause a destruction of the secreting wall and a cure of the disease.

I have before me the records of cases which have fallen under my own care, in which this ranula was very large. One was in a female child, aged only 5 years. The tumour was of the size of a large egg, and was only of three weeks' growth. The child was unable to articulate, and could not eat; the tongue being pushed quite out of its place against the roof of the mouth. A seton was put in, and the secretion let out; two ounces of thick albuminous fluid escaping. After one week, the seton was removed, and recovery followed.

In a second case, of a boy aged 8, the ranula was so large, that it appeared as a fluctuating tumour beneath the jaw. The tongue was also much displaced, rendering articulation difficult and indistinct. By the introduction of a seton, a complete cure was readily obtained.

It is scarcely necessary to dwell longer upon this subject; the special point of interest in these cases being their rapidity of growth, and the readiness with which a cure can be obtained by the use of a seton.

Warty Growth upon the Tongue. To illustrate this subject, I propose to quote a case of warty growth upon the tongue, which came under my care in 1857. It was in a female infant, aged 1 year; and the mother had observed the growth for five months. It occupied a greater part of half the tongue; the extreme tip was, however, free. The warts were raised and pedunculated. The child's health was otherwise good. A free application of nitrate of silver proved sufficient to destroy these growths, and recovery followed.

In contrast to this simple affection, I must quote a second case of warty growth upon the tongue; which was, however, due to a different cause, evidently constitutional. It was in a female child, aged

1 year, who had, one month after birth, shown every symptom of congenital syphilis. It had had the snuffles, and a general cutaneous eruption, which had been cured at St. Thomas's Hospital. The child remained apparently well for five months, when there was observed upon the tongue a warty growth, which gradually increased. When I saw the child, this growth was of about the size of a sixpence, and was precisely analogous to the condylomata which are found in the female genitals. A like growth also existed upon the lips and anus. Grey powder in one-grain doses, combined with seven grains of dried soda, was given twice a day; and in one month the child was well.

These two cases form an interesting contrast— one of simple warty growth, the second of syphilitic.

I will now proceed to quote a third case of warty growth of the tongue, of a totally different nature from either of the former cases. In its nature it is so distinct, that it is hardly possible for any one to fail to recognise it when once aware of the class of cases to which it belongs. I allude to a degenerating nævus. It was in a child 10 years of age, who had had a swelling in the tongue from an early date; but it was only during the last twelve months that it had put on the appearances which it presented when the child was brought to me. The posterior part of the tongue was covered with the peculiar vesicular and warty growth which is so characteristic of a degenerating nævus; and, without the history, the true nature of the case could be made out. But the mother had observed that the tongue in previous years was at times much swollen, particularly when the child cried; this fact being very visible, as the child used constantly to protrude the organ involuntarily when screaming hard. This temporary enlargement would rapidly subside, and the tongue again appear natural as to size. The above history clearly pointed to a vascular tumour or nævus.

With these three cases, the diseases of the tongue may be passed over; and I have been tempted to bring them before you mainly for their own interest.

I must, however, add one word respecting

Wounds of the tongue, and more particularly relating to their treatment; being anxious to show the necessity of treating all such cases by means of sutures, if the substance of the organ be much involved; for such wounds have a great tendency to gape, and much serious inconvenience may subsequently arise from a neglect of this measure. If they be treated, however, by sutures, good union will generally take place; repair here going on very rapidly. In a case which came under my care eight years ago, the evils of a neglect of this practice were well shown. It was in a girl aged 9 years, who had received an extensive wound of the tongue from a fall six months previously. The case had been neglected, and no surgical treatment had been employed. As a consequence, the right half of the tongue was nearly separated about the centre, and it was with difficulty that the child could speak or masticate. I pared the edges of the wound, and applied sutures; a good recovery taking place.

The *Tonsils* and their diseases hardly require any special observations. Their chronic enlargement may be found in children as well as in adults, although perhaps such cases are more common in early life. It is a disease essentially of debility, and is to be met by iodine and tonics; quinine and the iodide of iron being undoubtedly the best. If they become very large, indurated and pale, so as to affect the powers of deglutition and respiration, excision should be performed; the guillotine being the safest instrument to employ. In one case which came under my notice, these glands were so large as to prevent the patient's deglutition of solid food. The child was $3\frac{1}{2}$ years old, and had lived for six months upon liquids only.

I have never seen any good result from the local application of nitrate of silver, iodine, etc., in these cases; and have sometimes thought that they do harm. I never use them. Internal remedies prove generally successful.

Surgical Affections of the Anus and Rectum. We will now pass downwards towards the terminal end of the digestive canal, and consider if there be any surgical diseases of the anus or rectum which are either special to childhood, or are more frequently found then than during adult life; and the first to which I will draw your attention is *Polypus Recti*. This disease is by no means an uncommon one in children, but in adults it is without question comparatively rare. It has been only of late years that much attention has been paid to this subject.

In 1859, I had the pleasure of bringing before your notice a short paper, in which I wished to prove, that in children hæmorrhage from the bowel is a very certain indication of the existence of a polypus; and I brought forward at the time, to demonstrate the fact, many cases as well as preparations of the polypi which I had removed. Since that date, many similar cases have fallen under my care; and I must confess that it is a rare thing for me to meet with any instance of hæmorrhage from the bowel in children, which is not to be explained by the presence of such a growth.

These growths are generally found in children under ten years of age. In some cases, the discharge of blood from the bowel is constant, and the patient will be brought with its clothes stained, and its buttocks smeared with a bloody mucus. In these instances, the polypus will generally be found to be within, if not protruding from, the sphincter. In other examples, occasional discharges of blood will be observed, although not to a great extent, and will generally accompany and follow the act of defæcation. In others, again, the hæmorrhage will take place independently of any such process. In a patient exhibiting any of these symptoms, a careful local examination should be made. Much care is required in the examination, as the growth is readily passed by and overlooked. The best plan is to sweep the finger, passed well into the rectum, completely round the walls of the bowel. The polypus will thus be dragged from its attachment, and its pedicle will be made tense; thus arresting the attention of the examiner. By a careless examination it is almost sure to be overlooked, unless very large.

When it is once detected, the cure is not difficult. The removal of the growth is the only correct treatment; and this may, in a large proportion of cases, be readily done by simply hooking the finger over the pedicle, and breaking it off. I have never known any evil result from this practice, not even hæmorrhage; the process of lacerating the pedicle preventing this occurrence.

In other instances, the polypus may be brought externally, and ligatured. If the pedicle be thick, and the polypus high up, it should be dragged down by means of forceps, or by a wire noose, and then ligatured. The removal by the finger is the readiest and best practice, and it is the one which I have almost always followed.

After the removal of the growth, a cure may confidently be predicted; some care being always taken by the surgeon to satisfy himself that a second growth does not exist.

The following are brief notes of some of the more recent cases which I have treated:

Case I. J. P., aged $4\frac{1}{2}$, was brought to me after he had been losing blood at stool constantly for seven months, and at times even when walking. Nothing however protruded from the anus; and he was free from pain. Upon examination, a polypus was found, which was brought down by means of forceps, and ligatured; the growth being cut off below the ligature. Recovery followed.

Case II. A boy, aged 10, was brought to me for bleeding from the bowel of twelve months' duration; and for three months it had been profuse and constant. When seen, the child was blanched and powerless. A polypus of the rectum was at once detected, and torn away, convalescence rapidly following.

Case III. A boy, aged 7, for two years had been the subject of hæmorrhage from the bowel when at stool, and for one year had lost blood even when walking. Powerful straining had also been observed when at stool. On examination, a large polypus was found, with a pedicle which allowed the growth to pass downwards within the sphincter; this fact explaining the straining and the occurrence of the hæmorrhage which took place independently of defæcation. A ligature was applied to the pedicle, and the growth excised. Recovery followed.

Prolapsus Recti is another surgical affection which is common to both children and adults, but it possesses some peculiarities when found in young life which demand our notice. It is by no means a rare affection; but, in the majority of cases, it is only a symptom of some other disease—a result, and not a cause. It is constantly found in children who are the subjects of gravel or of calculus in the bladder, or of any disease causing irritability of the urinary organs, such as phimosis and inflammatory affections of the mucous membrane covering the glans penis, with adherent prepuce; it accompanies also constipation of the bowels, and the presence of worms. In whooping-cough, and in cases of great debility, it is also not unfrequent.

A correct interpretation of its cause is, therefore, plainly an important point to be acquired before any hope of its successful treatment can be entertained. It must be observed, however, that in the majority of cases of so-called prolapsus recti, there can be no doubt that the projection is only of the mucous membrane; but, in some rare and extreme instances, the whole bowel appears to be forced downwards, giving rise to a form of intussusception.

It is needless to quote examples illustrating all the various causes of prolapsus recti, for the experience of most surgeons will readily suggest many; but, in every case which comes before our notice, it is as well to examine carefully the condition of the urinary organs; to see that no stone exists in the bladder, and more particularly that there are no adhesions of the prepuce to the glans penis, or retained secretion from Tyson's glands; these concretions, without doubt, being amply sufficient, as well as the most frequent causes of all the symptoms.

If habitual constipation be the cause, it must be treated. I have seen a girl, aged 16, who for five years had been the subject of prolapsus recti whenever any action of the bowels took place, this action seldom being under five, six, or seven days. I may, in passing, remark upon the great benefit which patients suffering from this disease experience from the habit of relieving the bowels at night; that is, of assuming the recumbent position after the act. In these cases, the benefit is most marked; but the habit is a good one in all cases of rectal disease.

As a symptom of ascarides, prolapsus is very common; for this, a good calomel purge is often amply sufficient to obtain a cure; but, otherwise, enemata of some bitter infusion, such as quassia, will answer the same purpose.

In some cases, however, this disease is due to a weakness of the parts and a general want of constitutional power. In such instances, the general health must be attended to, and the bowel reduced after each prolapse. In many examples, this treatment answers every purpose; but in some, more energetic local treatment seems called for; and in these I have found much benefit and no evil result from the local application of the nitrate of silver to the whole surface; this must be done, however, rapidly and regularly, and the whole prolapsed membrane should be well covered, dried, and returned. It is seldom that more than one application is called for; and I have never seen any evil result from the practice.

Hæmorrhoids in childhood are a remarkably rare affection. I have seen many cases so called, but very few which were genuine cases of this disease; the majority of such being undoubtedly examples of rectal polypi. In looking down my notes of cases of polypi, I find the observation "treated for piles," to nearly all.

Anal Abscess and Fistula in Ano in early life are very rare diseases; that is, when compared with their frequency in adults. I have had one case under care in a boy aged 5 years; but this is the youngest subject in which I have seen this affection. They call upon me for no remark.

THE DIFFERENCES IN THE SURGICAL DISEASES OF THE URINO-GENITAL ORGANS IN THE CHILD AND ADULT.

ALMOST every surgical affection of the urinary organs in children is primarily manifested by irritability of the bladder. This symptom may exist alone, or it may be associated with others; but it is unhesitatingly the chief and most prominent symptom by which the attention of the parent is first directed to the part.

By far the most common cause of this irritability is to be found in the condition of the penis; a long prepuce, retention of the secretion from Tyson's glands around the corona, or an adhesion between the prepuce and the glans penis, being amply sufficient to produce this symptom, and, indeed, almost any or every other symptom of vesical disease.

I have seen this simple condition of penis produce every degree of irritability of bladder, even to hæmaturia. I have seen retention of urine result from the same cause, and also constant priapism.

The following cases will, perhaps, best tend to illustrate these points.

CASE I. A boy, aged 1½ years, was brought to me in July, 1862, with great irritability of bladder, which had been observed for many months, and hæmaturia of six weeks' standing; the pain which the child experienced during micturition was apparently very severe, as it was always followed by a shrill scream. On examining the penis, the prepuce was found to be very long and adherent to the glans. Circumcision was performed, and the adhesions were torn away, perfect recovery rapidly following.

CASE II. A boy, aged 3 years, was brought to me in Oct., 1862, his mother having observed that the child had constantly from birth played with his penis after micturition; and for six months he had at intervals a somewhat profuse hæmaturia. On examination, I found that he had an elongated and adherent prepuce. Circumcision was consequently performed with complete success, every symptom of distress rapidly disappearing.

CASE III. A boy aged 20 months was brought to me in June, 1861, with retention of urine; the mother stated that he had had no relief for three days, and he was becoming drowsy. He had been suffering from occasional attacks of retention of urine for six weeks, having passed two days on several occasions without relief. On examination, his prepuce was found to be firmly adherent to the glans penis, and a quantity of retained secretion was also present. Circumcision of the prepuce and its complete separation from the glans was performed; and a catheter was passed, by which at least a pint of urine was drawn off. Convalescence followed; and, although he was under observation for three months, no return of his symptoms had been observed.

CASE IV. A boy, aged 2 years, was brought to me in May, 1862, for constant priapism, the mother having observed that such a condition was always present. On examination, it was seen that he had an adherent and lengthened prepuce; consequently he was circumcised, and the adhesions were broken down, a complete cure following. It must be added, that this child was said to have had an attack of tetanus when teething.

It is hardly necessary to quote other examples illustrating the same subject. The above are typical of certain classes, and have been selected from their extreme severity; it is enough, therefore, to add the caution that, in all cases of irritability of the bladder, or in any cases in which urinary disorder is suspected, the first inquiry should invariably be directed to the condition of the penis; for I have shown that almost every symptom of urinary disease may be produced by an elongated or adherent prepuce.

When any young subject is brought to me with a history of calculus vesicæ, I am always at first disposed to doubt the fact of its existence; for, in by far the larger proportion of such cases, the symptoms will be found to be due to the simple affection I have just described; and too often an opinion has been given to a parent that a stone exists, when every symptom has been produced by the condition of penis which has been already dwelt upon. The importance, therefore, of recognising the truth of these remarks cannot be over estimated; for, if the nature of the case be mistaken, the fears of the parents of a child will be needlessly excited, the necessary and simple means which are required to effect a cure will be omitted, and the suffering of the child consequently prolonged.

Treatment. In all these cases, circumcision is the only correct practice. The prepuce must also be carefully separated from the glans penis; and all retained secretion from Tyson's glands removed. Common cleanliness is all that is subsequently demanded, and a good and permanent cure may, with confidence, be promised. No other measures will have a chance of succeeding; and it is only for the practitioner to realise the connection between the symptoms and their cause, to understand the importance and necessity of the practice which has been brought before you.

Idiopathic or Inflammatory Stricture is an unknown disease in early life; although cases of stricture are occasionally met with in practice, the nature of which is somewhat inexplicable; for, unless some injury to the urethra have been previously experienced, and as a consequence some subsequent contraction have taken place, producing what is described as a traumatic stricture, no other cause for an urethral contraction is generally looked for.

It is, perhaps, hardly fair to regard all cases of stricture which cannot be put down to the result of injury, as cases of irregular development; but such an explanation appears most probable, although I have no proof to bring forward in support of this theory.

I have had a case of so-called stricture under my care, in a boy only 9 years old; but I do not think it necessary to take up your time by dwelling at greater length upon the subject.

Extravasation of Urine. When extravasation of urine is found in the *adult*, it is almost invariably due to an urethral stricture; when found in the *child*, it is almost always produced by an impacted urethral calculus; for I have already stated that, independently of some accident to the urethra, idiopathic stricture is an almost unknown disease in young life; and I think it can be stated with equal truth that extravasation of urine from an impacted calculus in the urethra of an adult is very rare.

Extravasation of urine, therefore, when found in children, is almost always the result of an impacted calculus. It is to be treated also upon only one plan, and that is, by opening the urethra and removing the stone. It is not in every case that a calculus will be detected; for the stone may be small, and under such circumstances it may be lost in the sloughing and œdematous tissues; but in the majority of instances it will be detected at the time; and, if otherwise, it will be found at some future period in the discharges which come away.

I possess the records of at least a dozen instances of this affection, and in all of these the practice just laid down was followed; in such cases, also, a good result may be anticipated if they be taken early and the extravasation of urine have not been too extensive.

This point of difference in the cause of an extravasation of urine in the child and the adult is very marked, and is worthy of attention.

Stone in the Bladder. The two great points of difference between this disease as occurring in children and in adults, are the greater frequency of the disease, and also the much greater success of an operation in the former as compared with the latter, for it is a well established fact that two-thirds of the cases which come under observation in hospital practice are in children, and that not one in twenty of these cases proves fatal after operation. On the other hand, one-third only of the cases of stone in hospital practice occur in adults; and of these one-sixth die after operation when subjected to it during middle age, and one-half, at a later period. Again, in the majority of cases of stone in children, the general health of the patient is good; indeed, when compared with other children, the inmates of a hospital, the subjects of stone are by far the most ruddy and robust; and it is tolerably certain that the deposition of a calculus is not connected with any cachexia or feeble condition of the vital powers. It is only when the calculus has existed for a lengthened period that the general health of the child begins to fail, and danger to life from its presence and from its operative treatment is to be dreaded. In another place (*Medico-Chirurgical Transactions*, vol. xlv.), I have been enabled to show that in the majority of cases of death after lithotomy, whether in children or in adults, extensive renal disease is the unquestionable cause; and that this disease is a necessary result of the long continued presence of a vesical calculus.

In children, stone in the bladder is not a fatal

disease, unless neglected, and it is very amenable to treatment, the operation for its removal being most successful, not more than one case in twenty proving fatal. This mortality includes the special risks which attend lithotomy as every difficult operation; and also the more serious dangers to which a patient is always liable; renal disease being the chief. As a conclusion, therefore, I believe it may be left to the fancy or fashion of the operator to select the form of operation which may be proper. The lateral and the median appear to be at least equally successful when skilfully executed; the success of the operation depending little upon the one which is selected. If the patient be a healthy one, and either operation be carried out with caution, a good result may be fairly predicted; but if the child be cachectic and the stone have existed for some years, or for a large proportion of the child's life, however well either operation may be performed, a bad result may be anticipated; for it has been shown that, in nearly every fatal instance of lithotomy, renal disease is the true cause of death, and that without such a complication death is by no means common. The special risks of the operation itself and the results of accident are, of course, omitted in this estimate; but, happily, these are rare.

From the influence of example, practice, and preference, I always perform the lateral operation; and the straight staff is the one on which my choice falls; but these points, I imagine, are quite unimportant if the execution of the operation be skilful and correct, and renal complications be not present. So let the advocates of Allarton's or the median operation dispute the value of the lateral; and the advocates of the latter doubt the advantages of the median. For my part, I believe that both are right if rightly executed; and that a fatal result depends upon other and less doubtful grounds.

END OF SECOND LECTURE.

LECTURE III.

Mr. President and Gentlemen,—In the lectures I have already had the honour of delivering, I endeavoured to bring before you the chief points of difference between the surgical affections of the child and adult, as they are observed in the nervous, respiratory, circulatory, and urino-genital systems. I propose on the present occasion to point out in what way the diseases of the bones and joints differ in the child from those in the adult, and then to pass on to consider, as far as time will allow, the subject of tumours.

SURGICAL AFFECTIONS OF THE OSSEOUS SYSTEM IN CHILDREN.

Between the bones of children and adults there is one great anatomical difference, which it is important to bear constantly in mind while considering the pathology of the osseous system; and it is this— that in the *adult* a bone may be regarded as a separate thing, composed entirely of one piece; whilst in the *child* we must remember that it is made up of several parts—a central body or shaft, epiphyses or extremities, and the intervening soft and pulpy structure through which these grow and subsequently unite. In the adult, this union is supposed to have taken place, although it is well known that in some bones a perfect junction of the shafts with their epiphyses is not completed till a much later period of life than it is in others.

When considering the diseases of these structures in children, it is, therefore, always necessary to remember the fact that the bones are composed of separate portions; that these have distinct sources of nutritive supply; and that each is liable to be affected by disease as an independent part.

An epiphysis of a bone may be the subject of a disease which will not involve the shaft to which it is to be subsequently connected; and the shaft may likewise be similarly affected, and yet the epiphyses be left completely free. Disease may also attack the intervening soft and pulpy layer by which the growth of the shaft is carried on, and its subsequent union to the epiphysis is to be finally secured. In children, therefore, the diseases of the shafts and epiphyses demand a separate consideration; and the affections of the intervening pulpy layer will likewise receive an independent notice.

It may be remarked, that authors have failed to enlarge upon this great difference between the diseases of the bones of children and of the adult; and it is with some pleasure that I now bring the subject before your notice.

Inflammatory Diseases of the Shafts of the Long Bones in children are unfortunately very common; and, unless attributable to some external injury, the majority of cases are found in the cachectic and so-called strumous subjects. They are generally slow in their progress and certain in their results, the end being far too frequently a necrosis or death of the part. In idiopathic cases, the progress of this inflammation of the shafts is generally remarkably slow, and it is often marked by a torpidity and want of definiteness in the symptoms, which disarms the suspicions of a parent, causing the affection to be regarded with indifference; and this feeling produces a dilatoriness in seeking surgical advice, which has nothing but an injurious influence upon the ultimate result of the case.

In some instances, this disease is only manifested externally by a gradual expansion or apparent dilatation of the bone, the external and soft parts around appearing pale and natural; and it is not till many months have perhaps elapsed that other external signs of mischief make their appearance. If the disease, however, be allowed to proceed unchecked, death of the bone will to a certainty take place, when inflammation of the integument and suppuration will necessarily make its appearance, ending in the formation of sinuses and a cloaca communicating with the diseased bone.

The constitutional symptoms during the progress of these changes vary in their intensity in different patients. In the majority, they are very mild; little but an aching pain at the spot demanding notice. If the surgeon be consulted at this second stage of the disease, he will observe a marked difference in the temperature of the part, the integument over the diseased bone feeling much hotter to the hand than that around; and firm pressure over the bone will, in all probability, cause increased suffering.

On a careful examination of the bone at this time, it will be found generally enlarged, and apparently expanded, its external surface being smooth and uniform. At a later stage of the disorder, when the soft parts have become implicated and the bone has died or become necrotic, the external or visible symptoms are too well marked and positive in their nature to demand notice.

It must also be remarked that, in these cases of idiopathic ostitis, it is rare for one bone alone to be the subject of disease. The symmetrical bones, as a rule, sooner or later become more or less involved, and in some cases nearly all the long bones in the body; this fact pointing to the constitutional origin of the disease. I have had under my care for some time past a girl aged 11, who has been the subject of osteal disease for four years. The humerus, ulna, and radius of both arms have been diseased; and in both ulnæ exfoliation has taken place. Both tibiæ and fibulæ are now affected, with the right femur; and it is to be remarked that the left tibia is exactly three-quarters of an inch longer than the right; one bone measuring 11¾ inches, and the other 11 inches. During the whole progress of this case, a constant aching of the bones was the only symptom; and at times this was very slight, at others being very severe. Under the influence of tonics and rest, marked benefit has been bestowed; the diseased action in many of the bones has been arrested; and their gradual restoration to their natural dimensions has been observed.

This case also illustrates another point which is generally to be noticed in like instances; and that is, when the shafts of bones are diseased in children, the epiphyses are left sound; and that, even in such a typical case of the disease as the one I have just quoted, no epiphysis seemed to be affected in the slightest degree. This pathological fact tends of itself to prove the anatomical distinction between the parts; and when we proceed to consider the diseases of the epiphyses, a like point will have to be recorded.

The *elongation of an inflamed bone* is a pathological fact of extreme interest; it has been observed before, and noticed by different authors; but cases illustrating the point are not common. In the example quoted, the difference between the two tibiæ was three-quarters of an inch, and both were inflamed; the longer bone, however, being the worst; proving that, under the influence of the inflammatory process, a long bone increases lengthways, as well as in its diameter; although it may be a question whether, on the cure of the disease, the bone will regain its normal length, as it unquestionably may its normal dimensions in all other respects.

In the *treatment* of these cases there is but one principle to follow—to improve the general health. The disease is one essentially of debility, and is to be met by good living, good air, and tonics, as quinine, iron, cod-liver oil, or any other which may best suit the peculiarity of the case. The syrup of the iodine of iron is a good preparation, and is a favourite of mine; but any, or all in order, may be given with advantage. The inflamed limb must, however, be well raised and kept at rest, and warm fomentations afford much relief in the early stages. I have never seen any benefit result from the local application of iodine: it causes much pain to the patient by inflaming and irritating the integument; and seems to be quite useless. A warm fomentation gives much greater comfort, and, when combined with rest, is the most serviceable.

When the bone is necrotic, it is to be removed; but surgeons should be careful not to interfere in these cases too soon. The powers of nature, when fairly assisted, in restoring diseased parts in children to their normal condition, are almost unlimited; and the surgeon has, therefore, only to remove diseased bone when it is evidently lying alone, separated from its matrix, and acting, therefore, as a foreign body.

Too early attempts to interfere are followed by nothing but harm, and should therefore be avoided. Amputation of a limb for such an affection is rarely, if ever, to be entertained.

The necessity of an early detection of this disease is a matter of primary importance. In a large number of cases, the surgeon is not consulted till it has existed for a lengthened period, and has progressed to an advanced condition ; the mildness of the symptoms masking its true nature. Any aching of the bone, if constant, is a symptom of great value, for it is undoubtedly the earliest by which this cachectic form of ostitis is ushered in ; and such a symptom should never, therefore, be treated lightly ; for at this early stage of the disease, by rest and tonics, a speedy convalescence may with some certainty be predicted, months of misery be probably saved, and subsequent deformity prevented.

I might quote many cases to illustrate every detail in this matter, but it is hardly necessary ; the limits of my lectures urging me to pass on to another subject no less important—namely, the inflammatory affections of the *epiphyses*.

Inflammation of the Epiphyses or the Articular Extremities of Bones. The epiphyses of the bones of children are remarkably prone to inflammatory affections ; and it is due to this fact that diseases of the joints at an early age are so common and so obstinate. It may without fear be asserted that at least two-thirds of the cases of joint-disease which are found in children are due to inflammatory affection of the epiphyses; and the symptoms by which this disease is to be recognised are not obscure.

As previously stated, these epiphyses may be regarded as independent parts; they are still distinct as bony centres, and have not yet united to the shaft and become part of the body of the bone. As a result, the local symptoms which accompany these inflammatory affections are confined to the epiphyses, and seldom, even at a later stage, involve the shaft itself.

The earliest symptoms of any inflammatory affection of the epiphyses are somewhat obscure, a mere aching of the part being probably the only one which is to be noticed. To external observation at this early stage of the disease, there may not be any change in the appearance of the part, although some increase in the temperature of the integument over the bone is often to be recognised, when it is compared with the temperature of the integument above and below the seat of the disease. If the nature of the affection be not recognised, and some weeks, or, perhaps, months be allowed to pass without its detection, other changes will make their appearance; and of these the most marked is an absolute enlargement of the bone. It appears as a kind of dilatation and general expansion, and will be at once observed when a comparison is made between the corresponding epiphyses of the healthy and the diseased side. The bone at this stage will probably be somewhat more tender than natural, firm pressure over it exciting pain. The increase of heat in the part will at this period be also very palpable.

As time progresses, and if the disease be allowed to take its course, other symptoms will appear; and of these, some effusion into the articulation is the most positive, this being the result of a low form of synovitis, from the extension of the inflammation to the synovial capsule. The nature of the disease is, however, the same. The synovitis is clearly secondary to the ostitis of the head of the bone; and, if the disease in this part have not progressed too far, the synovitis will disappear when the ostitis is cured. It will be thus seen that the progress of the inflammatory affection of the epiphyses is precisely similar to the inflammation of the shaft; a gradual expansion of the bone, attended by an aching pain, forming the chief symptoms. At a later stage of the affection, also, the analogy still holds good; for, if the disease be allowed to go on unchecked, suppuration and the death of the bone will to a certainty take place; and, in the majority of cases, this suppuration will pass towards the joint, when disorganization of the joint-cavity will then be added to the difficulties of the case. It is an affection such as this which is often described as a strumous disease of a joint, but which is really essentially an ostitis attacking the epiphysis or head of a bone in a cachectic child, and is as free from any strumous or tuberculous disease as any other inflammatory affection of a cachectic type.

If the nature of this affection be detected at any early stage, a complete cure may unquestionably be obtained in a large proportion of the cases which fall under observation. It is too true, however, that in the majority of cases the disease is not recognised in its primary condition; the want of positive symptoms which are capable of exciting fear in the uneducated mind being the probable cause of this neglect, as the mere aching of the bone or joint is regarded only as a "growing pain," and the enlargement of the bone and increase of heat are not obvious enough to be observed. Should, however, the surgeon at this stage be consulted, he must be alive to the nature of the case. I always look upon a persistent "growing pain" with suspicion, and, on examination, seldom fail to find its cause; osteal disease being one of the chief. I would, therefore, wish to impress upon all men the necessity of carefully examining a limb the subject of such a complaint; and, in all cases of " growing pain," to fear some inflammatory affection of the osseous system. This fear may be wrong; but by carrying it into practice much benefit will be obtained, and many limbs and joints saved which would otherwise be lost. I could quote numerous cases illustrating these facts, both the evils of delay, and the benefits of the treatment I advise; but no practical benefit could be obtained by doing so, if I have succeeded in convincing you of the truth of these remarks.

Treatment. In the earliest stage of this form of disease, the treatment is very simple. Like the inflammation of the shafts of the bones, it is a constitutional affection, and one of debility. It is found in weakly and cachectic subjects, and in strumous or scrofulous, if a simple cachexia signifies the same. It requires, consequently, a tonic regimen and medical treatment; good air, good living, and tonics being absolutely essential. As local remedies, rest in the horizontal position and warm fomentations are likewise most important. Under such treatment, this disease may readily be checked in its earliest stage; that is, if no further organic change in the structure of the bone has taken place, than its infiltration with inflammatory product, and its consequent expansion.

When the inflammation has extended to the synovial membrane, and an effusion into the articulation has been produced, the same treatment is to be carried out as we have recommended in the less complicated example; but the prognosis of the case is not so favourable, and greater care in the carrying out of the instructions is to be observed. A successful issue is still, however, to be anticipated; a bad result being looked for only in the extremely cachectic and unhealthy subject. Should the inflammation, however, end in partial or complete necrosis or death of the bone, suppuration and disorganisation of the joint will, in all probability, be the result; but such cases will claim our notice under another head.

It will be remarked that, although this disease is an inflammatory one, leeching, blistering, and mercurials have not been alluded to; and I do so now simply to express the opinion that they are not only useless, but absolutely injurious. Cauterization of a joint affected with this disease is not a practice from which much benefit can be expected.

Inflammation of the Soft Pulpy Layer which exists between the Shafts of the long Bones and their Epiphyses. It is now well known by anatomists that, at the juncture of the shafts or centres of bones with their epiphyses, there exists a soft pulpy vascular layer of connective tissue, by which the growth of the body of the bone is maintained, and its subsequent union with the epiphyses secured. At an early period of life, this pulpy layer is well marked; for, as growth and development are at this time most active, the tissue is necessarily very vascular. Obediently also to the law that, in proportion to the activity of the vital process in the part, is its disposition to inflammatory disease, this tissue is unquestionably highly prone to inflammatory affections. I believe that a large proportion of the cases of acute suppuration about a joint in children have their origin at this seat; and this is often shown to be the case by an exfoliation of some portion of the surface of bone in contact with this pulpy layer. If the disease involve the whole connective tissue of the bone, a general exfoliation of the upper portion of the shaft may be produced; or, if this result do not take place, some arrest of development of the bone's growth may be the end.

The disease may make its appearance as an acute or a subacute affection. It is, however, generally acute, and is manifested by a marked swelling at the seat of the disorder, accompanied by great pain and constitutional disturbance. An abscess will then form, which may rapidly envelope the joint and upper part of the limb; and, when this has opened and discharged its contents, convalescence may result; or, what is more frequent, a piece of bone, varying in extent, will come away, and a cure follow.

The majority of cases such as I have described appear about the shoulder-joint, although the disease may be witnessed at any other articulation. The following cases may be given in illustration.

It must be added that, in this disease, the epiphysis itself is not often involved; this pulpy layer being more intimately connected with the shaft, the growth of which takes place through it; the epiphysis having an independent vitality and indedendent vascular supply. As a consequence, disease of this structure involves the shaft in preference to the epiphyses.

CASE I. Mary S., aged 8, applied to me in Dec., 1860, with a large abscess completely covering the whole shoulder. It had been coming for six weeks, and had caused intense pain and some constitutional disturbance. The abscess extended over the clavicle to the middle of the arm, and the slightest movement of the extremity aggravated the pain. A free incision was made into the part posteriorly, and a pint, at least, of pus taken away. A good poultice was then applied, and tonics given.

When I saw her again, a week afterwards, the swelling had entirely subsided; the joint could be moved without pain; and no signs of mischief within the cavity could be detected. The original wound was still discharging; and a second, over the bicipital groove, had made its appearance. The pain also had disappeared. It was clear, from this example, that the disease was not in the shoulder-joint, although it was not far from it. A probe, when introduced through the opening, passed upwards upon the bone, but no exposed bone could be detected. I believed the case at this time to be such an one as I have been describing. After the lapse of a few months, its true nature was revealed, for on April 28th, 1861, a piece of bone came away, which was evidently the upper portion of the shaft of the humerus, where covered with the pulpy vascular layer. On May 23, a second piece of the same character was removed from the anterior opening, and recovery ensued. I saw the child one year afterwards, and the cure was complete.

CASE II. A similar instance to the above appeared in the person of Mary B., aged 11, who came under my observation in November, 1862. Exfoliation of bone, which was evidently the upper surface of the shaft where in contact with the epiphysis, took place, and recovery ensued. The natural opening in this case was also over the bicipital groove.

CASE III. A third case occurred also in the person of George W., aged 14, from whom bone came away of the same nature, and recovery followed.

CASE IV. In Matilda D., aged $3\frac{1}{2}$, a like exfoliation was followed by a like result.

Other cases of the same nature as the preceding, might readily be extracted from my note-book, but they all tell the same tale—that of an acute or subacute abscess over the joint, followed by exfoliation of the layer of bone covered by the pulpy vascular tissue through which the bone grows, and subsequent recovery. It is to be remarked also that the abscess, if left to open naturally, bursts over the bicipital groove; this being the same position which an abscess from a suppurating shoulder-joint generally selects. All these cases have occurred in children, when the growth of bone and the activity of this pulpy layer is most active; and they, as a rule, have a good result. I have recorded in the *Guy's Hospital Reports* of the year 1862, an example of arrest of development of the humerus following an injury to the shoulder-joint, and I believe the case to be allied in its nature to those I have already quoted. In it, the accident was followed by an inflammation of this pulpy connective layer, and subsequently by an arrest of the growth and development of the bone; this growth being mainly effected through this tissue.

My colleague, Mr. Birkett, has also published in the same work a like case.

I have the notes also of a second case of arrest of development after suppuration and exfoliation of bone, in Emma C., aged 9 years. The abscess occurred when she was seven years old; and this was followed by exfoliation of bone, the cicatrix, when seen, being exactly over the head of the metacarpal bone of the left middle finger. This bone was also at least half an inch shorter than its neighbours. The movements of the joints were perfect, proving the integrity of the epiphysis, and indicating the probability that the original disease was situated in the soft pulpy layer, at the junction of the shaft with the epiphysis, that the upper portion of the shaft had come away, and had thus given rise to the subsequent deformity.

It is hardly necessary to dwell longer upon this subject, the few remarks which I have made being, I trust, sufficient to draw your attention to the point, and to mark the true nature of the cases which I have described, for they are all tolerably distinct.

I have thus briefly alluded to the three forms of

inflammatory affections which are found to attack the bones of children: namely, of the shafts, epiphyses, and intervening epiphysial structures; all of which are of an allied nature; but in each the symptoms of disease are quite distinct and readily distinguishable. I have also dwelt upon the anatomical causes of the differences between these osteal affections of early life and of the adult; and trust that I have not, in my efforts at condensation, left you uncertain as to my meaning, or as to the nature of the diseases I have described. I look upon the distinction which has been made as most important, and have, therefore, gladly seized this opportunity of bringing the subject under your notice, as I am not aware that the true nature of these cases has hitherto been described.

Fractures of Bones will now for a few minutes occupy our attention.

Congenital Fracture and Dislocation. The fact that congenital fractures of bone and dislocation occur in infants is now too well established to admit of a doubt, although the examples illustrating its truth are not abundant.

I have, however, seen but one example of intra-uterine fracture. This came under my care on Feb. 10th, 1862, the case having been brought to me by an intelligent pupil, in whose practice it occurred. The particulars of the case are as follows.

CASE. The mother of the child, two weeks prior to her confinement, fell heavily, and struck her abdomen sharply against the corner of a large weighing-machine. Signs of labour at once appeared, but passed off after rest; and, at the full period, after a natural labour, a female child was born, with a fracture of the middle of the right femur. When I saw the case two weeks subsequently, the existence of a fracture was quite clear, for it had not been fixed. The bone could also be readily bent into its natural position. I bound it up with card-board splints well padded with cotton wool, and kept it in position by broad strips of firm plaster. The fracture was rapidly repaired, and in one month convalescence was established.

The case is worthy of record for several reasons: first, from it being a congenital or intra-uterine fracture, an example of an accident which is recognised, but not often observed; secondly, to illustrate the facility with which the bone was restored to its normal position, a point which may always be observed in fractures of early life, from the fact of the bones being but partially developed, and containing more of the organic than inorganic elements; and thirdly, from the practice which was carried out, firm splints and hard appliances being rarely required in the treatment of such and allied fractures.

The limb should always be well protected from any unequal pressure by carded cotton, before the mill-board splints, which answer every purpose, are applied. Good, firm, linen strapping, closely surrounding the limb, binds the splints together more firmly and more securely than any bandage, and is far less cumbersome.

Respecting congenital dislocations, I have had no experience; but must refer you with pleasure to some excellent remarks upon this subject made by Mr. Athol Johnson, in his lecture published in the BRITISH MEDICAL JOURNAL for 1860.

It will be hardly necessary to detain you long with the consideration of ordinary cases of fracture of the bones; but it will be inconsistent with my original design, if I fail to point out to you the differences between the fractures of bone as found in children and in adults.

Fractures through the body of a healthy bone are not common in early life; by far the greater number of instances taking place in children who are the subject of rickets. In these cases, it is well known that the bones are readily fractured; and in the case represented by the drawing which I now hand round, the mother states that at least twenty different fractures have taken place. In such cases, repair of the fracture is however, rapidly performed, one month or five weeks being generally amply sufficient to complete the process.

Incomplete, or so-called "green-stick" fractures, form another peculiarity of infant life; the second name of "green-stick" explaining their true nature, the bone being fissured vertically into fibres, but not separated transversely, precisely in the same manner as a green stick. This form of fracture may be found in any long bone. I have seen it in the arm and forearm, and also in the thigh; but have never noticed it in the tibia or fibula.

The youngest patient in whom I have seen this injury was a boy aged one month. The fracture was in the centre of the humerus; it had taken place some time previously, and was supposed to have been produced at birth. The bone was readily restored to the right line and fixed in splints, a good recovery taking place.

CASE.—A very severe instance of this incomplete fracture presented itself before me in the person of a male child aged eighteen months. The right femur was really bent at a right angle, the foot and leg presenting outwards. It was produced by an attempt to catch the child when falling from the arms of its nurse. The fracture was readily restored by manipulation, and fixed with millboard splints, the limb being well padded with cotton wool, and the splints maintained *in situ* by strapping.

Fractures of the Clavicle are also very common. My colleague, Mr. Forster, in his practical work on *The Surgical Diseases of Children*, believes this fracture frequently to be one of the incomplete form; but as this bone is one of the earliest which is ossified, and at birth is, therefore, one of the strongest in the body, this opinion appears to me somewhat improbable. I would rather believe that these so-called fractures more often belong to the next class of cases to which I shall allude; viz., the separation of the epiphysis ; the body of the bone being separated from its sternal epiphysis.

This *Separation of a Bone at its Epiphysis*, from anatomical reasons, can only take place in early life. It is, therefore, a second point of difference between the cases of fracture of bones in the child and adult. It is a commoner form of injury at this time of life than a dislocation of a joint, and is, perhaps, more frequent than an ordinary fracture.

It is the most frequently found at the upper and lower part of the humerus and carpal extremity of the radius ; but it may take place in any bone. Considerable care is required to diagnose with correctness the nature of these accidents. They are too frequently looked upon as cases of dislocation, and are, therefore, maltreated ; the absence of distinct crepitus being the principal cause of this difficulty. They are to be treated in the same way as a case of fracture ; the epiphyses are to be restored to their normal position by manipulation, and splints applied, great care being observed that there is no undue pressure over the parts.

This leads me to say one word on the subject of dislocations, which are unquestionably cases of some rarity ; the majority of the so-called cases being a separation or dislocation of the epiphyses, to which we have just alluded. The treatment of these injuries, however, is the same in the child as in the adult.

ON DISEASES OF THE JOINTS IN CHILDREN.

ON such a subject, the whole course of three lectures might readily be expended, and to touch upon it in the third lecture may to some appear foolish; but, to be consistent, I shall confine my observations to the differences which are met with in practice between these diseases as witnessed in early and in adult life; and I shall condense my remarks within the narrowest possible limits.

To speak generally, it may with some confidence be asserted that there are but two structures entering into the formation of an articulation in which primary disease ever commences; viz., the bones and the synovial membrane. Primary disease of the articular cartilage is unknown; and I have shown elsewhere, some years since (*Diseases and Injuries of Joints*, 1859), that the diseases of the articular cartilage are secondary in the course of events, and are the necessary results of any lasting or acute affection of the articular extremities of the bones and synovial membrane. We may, therefore, exclude diseases of the articular cartilage from our consideration; at the same time recognising the fact that, in all extreme conditions of disease of the other two tissues, degeneration or so-called ulceration of this articular cartilage will take place, even to its total disorganization. Taking, therefore, the diseases of the bones and synovial membranes as the sole causes of primary disease of an articulation, I believe that I am strictly within the limits of truth when I assert that in adults at least *two-thirds* of all the cases of diseased joints commence in the synovial membrane, and *one-third* in the bone; whereas in children these proportions may be reversed; at least *two-thirds* of all the cases of joint-disease commencing in young subjects in the articular extremities of the bones, and hardly *one-third* in the synovial membrane.

This difference in the seat of the primary disease is the most important pathological distinction which I have to notice between the articular affections of early and adult life. I will not, therefore, run the risk of depreciating its importance by attempting an explanation of its cause, further than to add, that the more active growth and development of the bone of the child renders it more prone to inflammatory affections.

The next pathological fact to which I would briefly refer is the nature of the disease; as it is under the inflammatory affections that the majority of cases may be unquestionably classed. Inflammation of the articular extremities of the bones is, therefore, the cause of at least two-thirds of all articular disease in children; inflammation of the synovial membrane being the cause in the remaining third.

I would expunge entirely from our vocabulary the

terms strumous or scrofulous disease of an articulation. I know no positive condition to which such terms should be applied. They convey no definite meaning to my own mind, and I have never yet met with any one who could correctly understand their true nature. If by such a term it is meant to describe an inflammation of a part of a cachectic and feeble type, it may be intelligible; but surely, under these circumstances, the expression, "an unhealthy or cachectic type of inflammation," would be pathologically more correct, and would more readily convey the right pathological idea. "Unhealthy inflammation" is the term to which I affix my preference; and the phrase "scrofulous or strumous disease of a joint" I would not employ, unless I confined it to the cases in which true strumous deposit exists in a tissue; under which circumstances it would be rarely written.

In the earlier portion of this lecture, when dwelling upon the inflammation of the articular extremities of the bones, I pointed out the chain of symptoms by which the pathological changes in the bone are usually manifested. I showed how the earliest symptoms of disease in the bone were characterised by a gradual expansion of the epiphyses, accompanied by an aching pain and increased temperature of the part, with tenderness. In the second stage, when the inflammation had extended to other tissues, I showed how some enlargement of the articulation from effusion is the most prominent symptom; the synovial membrane having become secondarily involved, and the inflammation thus manifesting itself. If the disease should have progressed still further, and the inflammation in the bone have passed on to suppuration and destruction of tissue, the inflammation in the articulation becomes severe and of a destructive type; and it is at this period that the degeneration of the cartilage and a disorganisation of the joint will take place. In the majority of cases of joint-disease originating in the bone in children, this inflammatory process is of a low type and of slow progress; and, if unchecked, it has but one ending—the disorganization of the joint.

At the commencement of this disease of the bone, the joint is uninvolved, and is quite healthy; and, if rightly treated, no affection of the articulation may be produced. It is only in the later stages of the affection that the joint itself becomes involved; and this fact alone is enough to point out the danger of the case.

In the majority of cases of disease about the carpal and tarsal joints, unhealthy inflammation of the bone is the primary cause; and it is too well known that death of the bones involved is a frequent and destructive complication, the joints, as such, rapidly disappearing, and the bones exfoliating; and it rests alone with the constitutional power of the patient to determine the result.

Disease of the Synovial Membrane. It is unnecessary to dwell upon the symptoms which characterise synovitis, or to point out the differences between ordinary or healthy inflammation of the synovial membrane, and unhealthy inflammation which gives rise to pulpy or gelatiniform disease; they are essentially the same in children and in adults. In an early stage, both affections are remediable; but, if neglected, disorganization of the articulation will be the result.

Treatment of Joint-Disease. In the treatment of all cases of joint-disease, whether in the child or adult, it is right that we should have almost unbounded confidence in the reparative powers of nature to effect a cure, as it can be only through such a feeling that the surgeon would be preserved from interfering in the process of recovery, and from lapsing into the practice of that great error—meddlesome surgery. In the treatment of diseases of the joints in children, the necessity to bear this conviction in mind is far more necessary than it is even in adults; for it would almost appear as if there is no amount of mischief to the articulation of a child which is not remediable by natural processes; and that during young life every disorganised joint is capable of recovery, although perhaps it may be with anchylosis. I believe this opinion to be strictly true, with one exception; and that is, when any necrosed bone is present to keep up the irritation and prevent the completion of the cure. Under any other circumstances, a disorganised joint appears capable of repair; and, in the diseases of children, the proof of the accuracy of these opinions is constantly brought before us. The only drawback, however, to their universal application depends much upon the condition of the general health of the child; for it is in this that the grand fault always lies. The chronic inflammatory affections of an articulation are essentially diseases of debility; they are found in children who are naturally frail, or in others who have been brought into a cachectic condition by some other affection, such as fever, etc. Let these general conditions be improved, it is quite certain that the local affection will improve also; and that, as the general health of the child becomes re-established, the local health of the articulation will also return. In the treatment of joint-diseases, the surgeon's duty lies, therefore, in the carrying out of simple principles. He must, first of all, look to the general health, and employ all hygienic and medical means to improve and sustain the powers of his patients—good food, good air, and such tonic remedies as appear applicable, being of primary importance; for, without these, all local treatment will doubtless fail. Specific treatment, such as by mercurial remedies, is rarely required; and it would be better for such remedies to be altogether forgotten, than that they should be generally resorted to in chronic joint-affections. In some few instances they may be of service, but such cases are few and far between.

As to the local treatment of joint-affections, there is but one principle upon which we should always act; and that is, to remove all local sources of irritation by keeping the affected joint at rest, and by soothing pain. The first indication is best carried out by the application of a good splint to support the limb, leaving the joint uncovered for observation; and the second is well secured by the application of warm fomentations, either by flannel stupes or strips of lint, kept moist by warm water or poppy decoction. Counter-irritation in children is hardly ever required, and must be employed with great care and caution. It is very rarely indeed that I have recourse to such a practice, finding other treatment equally efficacious.

The above remarks apply more particularly to cases in which diseased action is still progressing, although they are applicable to others in which only the effects of disease remain; but, in this latter class of cases, some additional local treatment may be used with advantage; and it is in such that the benefits of pressure are well exemplified. Firm, equal, and steady pressure upon a joint by means of strapping, cannot be too highly extolled; for few things have greater influence in exciting the reabsorption of inflammatory deposit, and thus of restoring the parts to their normal condition.

After these remarks, it may not be unfairly asked if operative interference is ever called for in the treatment of joint-disease when taking place in early life? and to answer this question, I must occupy your attention for a few minutes.

As a general answer, it may without doubt be affirmed that far fewer cases of joint-disease require operative interference in children than in adults; for it must be again remarked, that the natural powers in the repair of mischief are almost unlimited in early life, and operative interference is, as a rule, not only unnecessary but mischievous. Again, it is only in extreme conditions of joint-disease that any operation is ever needed, and then only when the bone is extensively involved and has become necrotic; the suppuration being kept up, and the repair of the disease being prevented by the presence of the foreign body—dead bone; and, if this be removed, the subsequent repair of the part may with some confidence be predicted; it is, therefore, in these cases alone that any operation is called for on account of local conditions. It would be foreign to my purpose to enter into the question of the propriety of operative interference on account of the general failure of the child's health. In certain cases, this practice may be needful for higher motives than local considerations; inasmuch as the life of the child may be imperilled unless the removal of the source of irritation be carried out. But, regarding the case purely from local considerations, it is only in cases of necrosis of the articular extremities of a bone that operative interference is demanded; in all others, the subsequent cure of the case may with confidence be left to natural processes. It is not, however, in all cases of necrosis in a joint that an operation may be required, for, in many instances, the necrosed bone, if limited in extent, will come away, and recovery ensue. This fact was well illustrated in a case which came under my care in the year 1860.

CASE I. A boy, aged 5 years, had been the subject of diseased shoulder-joint for one year and a half; and for one year suppuration had existed. Partial anchylosis had also taken place. After he had been under observation for some months, a piece of bone came away, which was evidently part of the articulation, and recovery by anchylosis followed.

Instances like this also occur in adults, and I have met with several. In other cases, however, operative interference is unquestionably demanded; the diseased bone being either too extensive or too deeply placed for external exfoliation by natural processes. In these instances, its removal by operation is justifiable and essential; and the following case is a good illustration of the practice.

CASE II. A boy, Henry F., aged 5 years, came under my care in July, 1861, with a disease in the right hip-joint of two years' standing, suppuration in the part having been present for one year. It was quite clear that total disorganisation of the joint had taken place; and that dead bone existed, as it was readily felt upon passing a probe down the sinus situated behind the trochanter. Regarding the case as one of disease of the joint, secondary to inflammation and necrosis of some portion of the femur, I proposed an operation for the purpose of its removal; and on August 27th I carried out the practice. Upon making my incision behind the trochanter, a piece of dead bone within the hollow of the trochanter was at once observed; and, on further examination, it was found that the head of the femur had separated at its line of junction with the neck, and was lying loosely in the acetabulum. I therefore thought it best to excise the head and neck of the femur below the trochanter; and the preparation which I now show to you contains the specimen. I may add, that the acetabulum had been deprived of its cartilaginous covering, but was otherwise healthy. Convalescence rapidly followed; and in three months the child got up, and was able to stand upon and flex the limb without pain. The wound also had quite healed. He left the hospital and unfortunately contracted measles, which reduced his powers extremely; a large abscess also formed in the right thigh below the seat of the original disease. For this he was re-admitted; and, after good feeding and tonics, this gradually disappeared, and he left with a good, useful limb. He is now quite well. The limb is about one inch shorter than its fellow, but otherwise it is quite sound. The boy can walk and run freely without the assistance of any stick. He can stand upon the limb alone, and bend or rotate it in any position. Indeed, the result of the case is as satisfactory as could be wished.

This case is given as a good illustration of joint-disease, the result of inflammation and necrosis of a bone. Natural processes could hardly have proved sufficient to get rid of the sequestrum and exfoliated epiphysis; and an operation was therefore strongly called for. The success of the case was most satisfactory.

As further illustrations of nature's powers, assisted but feebly by art, in repairing extensive injury or disease of an articulation, the following cases may be read with interest. They are given as typical examples of distinct classes. The first was the result of injury.

CASE III. A boy, aged 10, was admitted under my care into Guy's Hospital, in July, 1859, with a compound dislocation of the right ankle-joint, produced by a fall of ten feet off a tree. The foot was twisted to a right angle with the leg; the sole presenting inwards: and the superior articular surface of the astragalus appearing through a large wound below the external malleolus. The articular surfaces also of the tibia and fibula were exposed to view. The fibula was broken three inches up. The dislocation was readily reduced by extension, and the foot fixed in position with a posterior and side interrupted splints; water-dressing being applied to the wound, and powdered ice in a bag around the foot. Everything went on well from the day of the boy's admission; and when he left the hospital, he was quite well, and what is more, had good movement in the articulation.

The success which followed the treatment of this case was most marked, although it is hardly to be expected that in all such severe injuries a good recovery with movement of the limb will follow; nevertheless, I believe, that in a large proportion of like cases in children, an equally good result may be anticipated; and of this I am certain, that it may be obtained far more frequently in children than in adult life.

I must give one other case as an illustration of the recovery of a disorganised joint from disease with movement, as typical of a second class of cases; and will select the ankle again.

CASE IV. A boy, aged 8, had been the subject of joint-disease for three years, and with suppuration for two. When he was admitted into Guy's under my care, the right ankle was riddled with sinuses; and a probe could be readily passed through the articulation. Exposed, but not dead bone, was also observed. The child's health was bad; and it was a question, when I first saw him, of amputation, as it was feared that the health of the patient was incapable of supporting the tax upon its powers. Tonics and good diet, with perfect rest of the limb, however, worked wonders. With the improvement of the general health of the child, the local disease took on a healthy action; and in six months all sinuses had healed, and diseased action had subsided. In another three months, all appearances of disease had vanished, and the child left the hospital well, with good movement of the articulation. I have seen the child since, one year after treatment, and the foot is as strong and sound as its fellow.

I might quote numerous other cases of disease of a joint to illustrate the truth of the remark which I made when first taking into consideration the subject of joint-disease—that in almost every instance, even the most severe, natural powers alone are amply sufficient to effect a cure. Time will not allow me to carry out these wishes; but I can fearlessly appeal to the experience of all surgeons to bear out the truth of the remark. Such experience, however, is of little use, if some practical deductions be not drawn to guide us in future practice; and I think that there can be little doubt as to the lesson which it teaches, for it is to leave Nature to herself. Let the surgeon aid to the utmost the natural powers by hygienic and general tonic means, and treat the local disease by placing it in the position in which it is least exposed to sources of irritation and local injury, relieving symptoms, and looking for indications of treatment, with the feeling of confidence that in time Nature's efforts will effect a cure even in apparently hopeless examples.

I must add one word upon the subject of excision of joints in children; and it will be to repeat what must have been gathered from the observations which I have already made—that it is rarely required. In necrosis of a joint, the operation for the removal of the dead bone may be called for, but such cases cannot with fairness be described as cases of excision, for they must be classed with other instances of removal of dead bone.

I have shown also that operative interference is rarely if ever demanded for any other local condition, and although I am prepared to believe that, in the joints of the upper extremity and hip, excision of the articulation may be called for, I am still, as I have always been, disposed to deny the value of the operation when applied to the knee; the weight of evidence which has recently been adduced certainly tending to confirm me in the truth of this opinion, in which I am now supported by many of the former advocates of excision, who appear doubtful of the value of the practice when applied to children.

ON TUMOURS IN CHILDREN.

The pathological nature of the tumours found in children differs but little from those in the adult.

They may possess peculiarities which are due to the rapid growth and structure of all infantile tissues; but, in their morbid and natural pathology, they are strictly analogous. There are the simple or innocent, and the malignant or cancerous tumours. In both forms, fibre-tissue is deficient, and cell-development abundant. Of the simple tumour we have, therefore, the fibro-plastic or fibro-cellular more frequently than the fibrous; and of the cancerous, the carcinoma medullare rather than the carcinoma fibrosum.

Tumours also, when they appear in any gland or tissue, as a rule have a rapid growth; the increase of the tumour keeping pace with the increase of the body; the cell-growth and multiplication of the abnormal equalling the cell-growth and development of the normal tissues.

In the bones of children there seems to be a special tendency to the development of tumours; cartilaginous growths appearing in these parts more frequently at an early period of life than at a late one. Cartilaginous or enchondromatous tumours of bone are almost always found in children; and, when in the adult, they have generally originated in early life. The explanation of this fact does not appear difficult, when we recollect that bony development almost always proceeds through the cartilaginous; for the cartilage-cell has but to remain as such, and to go on repeating itself, instead of progressing towards the development of a bone-structure, in order to form a tumour. This cartilage or enchondromatous tumour, under these circumstances, appears developed within a bone, and expands its shell. A most beautiful instance of this disease has recently come under my care; it was in a woman aged 32, and was of congenital growth, its increase having been gradual. It occurred in the little finger, and was treated by amputation. The specimen I now hand round.

Congenital Tumours. One of the most frequent forms of congenital tumours is the *sebaceous*, and its most common seat is the neighbourhood of the eyebrow. These sebaceous tumours have also one peculiarity; they are not situated in the skin, as are the other forms, but are placed beneath the muscle and close upon the bone. They very generally contain hair, in addition to the ordinary sebaceous matter; and this is found either in masses, or like fine eyelashes. These sebaceous tumours may appear also in other parts. I removed one from the labium of a child four months old, containing abundant hair; and a second, of the size of a walnut, from the extremity of the coccyx of a girl aged 10 years.

Congenital *fatty* tumours are not infrequent; and they occasionally possess a peculiarity for which it is difficult to account. In adults, fatty tumours are, as a rule, encysted; that is, they are distinct tumours, and can be enucleated. In children, an opposite condition is not infrequently found; the fatty tumour appearing as a diffused growth, and without distinct limits. I had an instance of this under my care in 1857, in a boy aged six weeks. He had a diffused congenital fatty tumour around the right elbow-joint, and this increased with his growth. It had no boundary, nor any borders; but its margins gradually disappeared in the tissues around. I have witnessed a second case in a child aged 5, a patient of Mr. Cock's; and in this instance the region of the occiput was the seat of the disease. I have also under my care at the present time a child aged one year and a half, with the same affection; the tumour occupying the left axilla and side of the chest.

Congenital *pedunculated fibro-cellular* tumours of the integument are very common, and need no desscription. The best example I have ever seen was in a male child who came under my care a few months ago, when only three days old, with a pedunculated tumour, of the size of a large walnut, growing behind the left ear; it was black from effused blood, and was very firm. It was excised, and recovery followed. The preparation and drawing, which I hand round, indicate its size and nature. An interesting case has also lately been under my care, of congenital fibro-plastic tumour growing from the little toe of a girl aged 8; it was of the size of a walnut, and occupied the whole of the extreme phalanx of the toe. It was amputated; and the specimen I now hand round.

Congenital *cancerous* growths also occur, but it has never fallen to my lot to witness an example.

Non-congenital Tumours: Sebaceous Tumours. One of the most singular cases of sebaceous tumour which has ever come under my care was in a female child aged five months, whose face and head were literally studded with sebaceous tumours, varying from the size of a hempseed to a large almond. They had been of three months' growth, and were in all stages of development; some just appearing, others suppurating; and some had enucleated their contents, and presented excavated granulating surfaces. I removed many by pressure with the nail, and the larger by incision; giving tonics, and maintaining cleanliness. Recovery followed.

The fatty and fibro-cellular and fibro-plastic tumours need not occupy our attention, as they present no points of contrast or peculiarity which deserve notice. But I cannot refrain from quoting a case of *simple cystic tumour* which came under my care in 1861. It was in a boy aged six weeks, and was, when seen, of about the size of a small egg, situated on the right side of the median line over the occiput. It was evidently cystic, and appeared to be connected solely with the soft tissues. It was observed the day after its birth, and was believed to have been congenital; and was then of the size of a small nut. I tapped it, drawing off some fluid stained with blood; and the cyst collapsed. When seen three weeks afterwards, no appearance of a re-collection could be observed. I have no later observation to record.

Keloid Tumours. These do not appear to be infrequent growths during young life. I have seen several such. One was in a male aged ten years. The keloid tumour was over the forehead, and of the size of a nut. It was said to have been congenital, and was growing. It was excised, and recovery followed. I have also observed the same disease in a boy six months old, growing from the cicatrix on a spot from which I had removed a nævus by ligature some months previously. The same disease appeared also in the child of a cousin of the former patient, in the cicatrix of a nævus which I had similarly treated.

Cancerous Growths are not uncommon in young life, and these appear to attack the bones more frequently than other parts. The best example which I have seen was in a boy aged twenty months. The tumour was in the hand, and occupied the position of the metacarpal bones, spreading the tendons over it. It was globular, as seen in the drawing and preparation; and was, as usual, of a medullary character. The glands were uninvolved. No other treatment than amputation could be thought of; and this I did in September, 1859. In 1862 the child was well, and had had no return.

Tumours of the Testes. Tumours of the testes must occupy our attention, although only for a few minutes. I have the records of one interesting case of inflammation of both organs in a boy six months old, who had had an enlargement of both glands from birth. When I first saw the child, the right was of the size of an unshelled almond, and the body of the testis was the part involved; while the left was about half the size of the right. I regarded the case as inflammatory, although no history of injury nor of syphilis could be obtained; and gave tonics. In one month the right testis suppurated and subsequently recovered; the left gradually became smaller.

I have also had under my care a child only six weeks old, with an abscess in the right testis of a day's duration, after inflammation for a week. No appearances of injury were visible, although, from the rapidity of the attack and progress of the disease, perforation with a pin appeared probable.

Respecting *cancerous disease*, I can relate an interesting case of a boy aged two years, who came under my care at Guy's Hospital in February, 1860, with medullary cancer of the right testis of six months' growth. The organ had enlarged gradually for four months, and for the last two had increased more rapidly. When excised, it was of about the size of a small bantam's egg. On April 3rd I removed the gland; and a speedy recovery followed. He remained well till May, 1862, nearly two years, when a return growth appeared at the original seat; and this increased rapidly for three months, when I again saw him. At this date there was a tumour of the size of a walnut in the position of the original testicle; it appeared moveable, and gave but little pain. No glandular enlargement could be detected. On August 6th, 1862, I again excised a portion of the growth, but found that a narrow neck of it dipped downwards into the perinæum, and could not be removed. The wound healed kindly, but the growth increased; and in November, 1862, when I last saw the child, he was sinking from apparently cancerous infiltration of the lumbar glands.

Strumous disease of the testes may also be found during young life. I have seen it in a child two years old, and of one year's growth; and again in a second, aged 2½, in which the disease had existed six months. In the last case, excision was called for; and the gland contained a large mass of strumous deposit, which was softening down. Recovery followed. Both of these cases were under my own care.

Inflammation of the Sterno-Mastoid Muscle. I must not omit to dwell upon a class of cases to which attention has not been much drawn, although it can hardly be doubted that most surgeons must have had such instances under their charge, although their true nature does not appear to have been understood. I allude to thickening and enlargement of the body of the sterno-cleido-mastoid muscle. I have observed it in three separate instances; and have always, from the ill-defined nature of the swelling, looked upon it as inflammatory.

CASE I. A boy, aged 3 months, was brought to me in 1857. The mother stated that the disease had appeared three weeks, and had gradually increased. The muscle was hard and much thickened; and, on manipulation, the child gave evidence of pain. I gave cod-liver oil and ordered warm water fomentation, which rapidly dispersed the disease.

CASE II. A female child, aged 2 months, came under my care in 1858. The disease was of one week's standing, and was of the same nature as in the last case. It was cured also by the same treatment.

CASE III. This case occurred in 1862, and was in a male child aged 5 months. The disease had existed five weeks, and was very marked; the body of the muscle stood out like a distinct tumour, but it had no defined boundary. Quinine and fomentation proved rapidly successful, and a cure resulted.

In none of these cases could any history of injury be traced. The disease was in all alike; it was in the sterno-mastoid muscle, and was evidently inflammatory; the inflammatory product apparently infiltrating the body of the muscle.

My colleague Dr. Wilks has recently reported several cases of a like kind; but, with these exceptions, I have no other record of the affection.

Tumours at the Umbilicus. The fleshy tubercle at the umbilicus is an affection which occasionally comes under notice; it is evidently composed of nothing more than exuberant granulations, as it always follows the separation of the umbilical cord, and is readily removed by the application of a ligature. I have had five cases under my care within the last few years in children, aged respectively fifteen, six, and five months, seven and five weeks; and in all a cure was at once effected by ligaturing the growth.

In connexion with this subject, I cannot refrain from quoting the brief note of an interesting case of growth from the umbilicus, of a totally different nature from those I have described. It was in a boy aged 8 years, who had a fleshy growth springing from the umbilicus, precisely like the glans penis of an infant. It was very red, and covered with mucus; and in its centre there was a distinct canal, through which urine passed; when the child retained its urine, it flowed freely from the opening; for it appeared tolerably certain that the canal was an open urachus. I was anxious to admit the child into Guy's, to see if anything could be done for his relief; but the mother refused, and I consequently had no opportunity of carrying out my wishes.

Cases like these are very rare, and are, therefore, worthy of a separate record.

I have thus, in these three lectures, brought before you the principal points of difference between the diseases of the various systems in the child and adult; and have, as far as I have been able, dwelt upon many of the special affections of early life. I have attempted to find an explanation of these differences, in the physiological and pathological processes as witnessed at these different periods of existence; and I have aimed at giving principles, rather than details of daily practice. How far I have succeeded I must leave to your kind judgment to decide. But of this I am sure, that I have advanced nothing which experience has not sanctioned; and, if I have been too short and brief in many of my descriptions, I have been so from the necessity which has been always before me of confining my observations within the few hours allotted to these Lettsomian Lectures. In conclusion, I have to express my thanks for the kind attention with which you have followed me; and to assure you that, if I have added anything either of interest or of fact to the important subject which has occupied our notice, I have been amply repaid, and shall feel that the object for which these lectures was instituted has not suffered from having been intrusted to my care.

THE END.

INDEX.

Ædema of larynx, 66.
Amussat's operation, 37.
Analysis of cases of hare-lip, 16.
Anus, malformations of, 26; surgical affections of, 87.
Anatomical differences in crania of children and adults, 57.
Arrest of development in bones, 53.

Bones, inflammatory diseases of their shafts, 103; of their articular extremities, 107.

Cancerous tumours, 140.
Cases of malformed anus, 34; of colotomy in right groin, 42; of deficient bones, 53; inflammation of pulpy layer at end of bones, 114; of irritable bladder, 93; of diseased joints, 130-134; of tumours, 137-140.
Circulatory system, differences between surgical affections of, in child and adult, 76.
Cleft palate, 25.
Clavicle, fractures of, 120.
Colotomy, 39; conclusions on, 50.
Congenital fractures and dislocations, 117.
Concluding remarks, 145.

Differences between physiology of child and adult, 3.
Digestive system, differences between affections of, in child and adult, 81.
Diagnosis of ostitis, 104.
Deaths in lithotomy, 98.
Dislocation, congenital, 119.

Elongation of inflamed bones, 105.

Enchondroma in children, 136.
Epiphyses, inflammation of; symptoms, 112.
Exploratory operation, its value in malformations of anus, 33.
Excision of nævi, 79.
Extravasation of urine, 96.

Fatty tumours, congenital, 137.
Foreign bodies in air-passages, 68; in nostrils, 73.
Fractured ribs in children, 75.
Fractures, congenital, 118.

Growing pains, 110.
Green-stick fractures, 119.

Hare-lip, 8; origin explained, 9; nature and extent of deformity, 10; its seat, 11; relative frequency in the sexes, 13; best age for operation, 15.
Hermaphrodism, 53.

Infantile diseases, greater activity of, 6.
Inguinal colotomy on *right* side advocated, 39.

Introductory remarks, 1.
Injuries to thorax, 74.
Irritable bladder, its causes in children, 93.

Joints, diseases of, 122.
 „ treatment, 127.
Keloid, 140.

Larynx, œdema of, 66.
Laceration of lung without fractured ribs, 74.
Lithotomy, median and lateral compared, 99; causes of death in, 98.

INDEX.

Littre's operation, 38.
Malformations of anus and lower bowel, 26; their classification and treatment, 26; pathology, 27.
Mechanical treatment of deformities, 55.

Nævi, 77; pathology of, 76; treatment of, 78.
Nervous system, differences between affections of in child and adult, 57.
Nostrils, foreign bodies in, 73.

Operation, best form of, in harelip, 19-22.
Orthopœdic surgery, 54.
Osseous system, differences of, in child and adult, 101.
Ostitis, early symptoms of, 103; treatment of, 106.

Polypi recti, 87; their frequency and diagnosis, 88.
Prolapsus recti, 90.
Pedunculated tumours, 138.

Ranula, 82; its treatment, 83.
Respiratory system, differences between surgical affections of, in child and adult, 65.

Salivary fistula, 81.
Sebaceous tumours, 137 and 139.

Specialities of surgical diseases of children, 8.
Statistics of hare-lip, 13; of stone in bladder, 98.
Stone in bladder, 97.
Stricture in children, rarity of, 96.
Sterno-mastoid muscle, inflammation of, 142.
Surgical affections of osseous system in children, 101.
Synovial membrane, diseases of, 125.

Tenotomy, value of, considered, 54.
Testes, tumours of, 140.
Tetanus in children, 62; its frequency after injury, 63.
Tongue, wounds of, 85; warty growths on, 83.
Tonsils diseases of, 86.
Tracheotomy, 69.
Tumours in children, 135; their classification, 136.

Urachus, open, 144.
Urino-genital organs, differences between those of child and adult, 92.
Umbilicus, tumours of, 143.

Warts on tongue, 83.
Wounds of „ 85.

GEO. P. BACON, PRINTER, LEWES.

www.ingramcontent.com/pod-product-compliance
Lightning Source LLC
Chambersburg PA
CBHW030318170426
43202CB00009B/1053